MR. JEFFERSON AND THE GIANT MOOSE

Mr. Jefferson

===== AND THE =====

GIANT MOOSE

★ ★ ★

NATURAL HISTORY IN EARLY AMERICA

Lee Alan Dugatkin

THE UNIVERSITY OF CHICAGO PRESS

Chicago and London

LEE ALAN DUGATKIN is professor of biology at the
University of Louisville. His books include *The Altruism
Equation* and *Cheating Monkeys and Citizen Bees: The
Nature of Cooperation in Animals and Humans.*

The University of Chicago Press, Chicago 60637
The University of Chicago Press, Ltd., London
© 2009 by The University of Chicago
All rights reserved. Published 2009
Printed in the United States of America

18 17 16 15 14 13 12 11 10 09   1 2 3 4 5

ISBN-13: 978-0-226-16914-9 (cloth)
ISBN-10: 0-226-16914-6 (cloth)

Dugatkin, Lee Alan, 1962–
Mr. Jefferson and the giant moose : natural history in early
America / Lee Alan Dugatkin.
p. cm.
Summary: Capturing the essence of the origin and evolution
of the so-called "degeneracy debates," over whether the flora
and fauna of America (including Native Americans) were
naturally weaker and feebler than species elsewhere in the world,
this book chronicles Thomas Jefferson's efforts to counter
French conceptions of American degeneracy, culminating
in his sending of a stuffed moose to Buffon.
Includes bibliographical references and index.
ISBN-13: 978-0-226-16914-9 (cloth : alk. paper)
ISBN-10: 0-226-16914-6 (cloth : alk. paper)
1. Degeneration—Environmental aspects—America—
Historiography—18th century. 2. Human beings—Effect of
environment on—America. 3. Life sciences—History—18th century.
4. Jefferson, Thomas, 1743–1826—Knowledge and learning.
5. Buffon, Georges Louis Leclerc, comte de, 1707–1788—Attitudes.
I. Title.
QH528.D843 2009
[B885.Z7]
508.097'09033—dc22
2009013816

For Aaron D., Fred C., and Lena P.

CONTENTS

A Moose More Precious Than You Can Imagine

Americans of the Revolutionary War era were understandably touchy about their standing compared with that of Europeans. It was one thing for the Europeans, particularly the French, to refer to Americans as upstarts, malcontents, and threats to the monarchy—in a sense many of them *were* all that. It was another matter entirely to suggest that all life forms in America were degenerate compared to those of the Old World. Yet that is precisely what Count Georges-Louis Leclerc Buffon, one of France's most distinguished Enlightenment thinkers, and one of the best-known names in Europe at the time, claimed.

In his massive encyclopedia of natural history, *Histoire Naturelle*, Buffon laid out what came to be called the theory of degeneracy. He argued that, as a result of living in a cold and wet climate, all species found in America were weak and feeble. What's more, any species imported into America for economic reasons would soon succumb to its new environment and produce lines of puny, feeble offspring. America, Buffon told his readers, is a land of swamps, where life putrefies and rots. And all of this from the pen of the preeminent natural historian of his century.

There was no escaping the pernicious effects of the American environment—not even for Native Americans. They too were degenerate. For Buffon, Indians were stupid, lazy savages. In a particularly emasculating swipe, he suggested that the genitalia of Indian males were small and withered—degenerate—for the very same reason that the people were stupid and lazy.

The environment and natural history had never before been used to make such sweeping claims, essentially damning an entire continent in the name of science. Buffon's American degeneracy hypothesis was quickly adopted and expanded by men such as the Abbé Raynal and the Abbé de Pauw, who believed that Buffon's theory did not go far enough. They went on to claim that the theory of degeneracy applied equally well to transplanted Europeans and their descendants in America. These ideas became mainstream enough that Raynal felt comfortable sponsoring a contest in France on whether the discovery of America had been beneficial or harmful to the human race.

Books on American degeneracy were popular, reproduced in multiple editions, and translated from French into a score of languages including German, Dutch, and English; they were the talk of the salons of Europe and the manor houses of America. And it wasn't just the intelligentsia of the age who were paying attention—this topic was discussed in newspapers, journals, poems, and schoolbooks.

Thomas Jefferson understood the seriousness of Buffon's accusations, and he would have none of it. He was convinced that the data Buffon and his supporters relied upon was flawed, and possibly even intentionally so. And Jefferson quickly realized the long-term consequences, should the theory of degeneracy take hold. Why would Europeans trade with America, or immigrate to the New World, if Buffon and his followers were correct? Indeed, some very powerful people were already employing the degeneracy argument to stop immigration to America. What's more, this insipid theory challenged the entire premise of the American Revolution: that man could rise to any heights for which he worked.

Jefferson led a full-scale assault against Buffon's theory of degeneracy to insure that these things wouldn't happen. He devoted the largest section of the only book he ever wrote—*Notes on the State of Virginia*—to systematically debunking Buffon's degeneracy theory, taking special pride in defending American Indians from such pernicious claims. The author of the Declaration of Independence employed more than his rhetorical skills in *Notes*. Jefferson produced table after table of data that he had compiled, supporting his contentions. Parts of Jefferson's book were reprinted in dozens of newspapers across the United States in the 1780s. Even a hundred years after that, one Jefferson scholar called *Notes on the State of Virginia* arguably the most frequently reprinted Southern book ever produced in the United States to that time.

As minister to France, Jefferson knew Buffon, and even dined with him on occasion. He was confident that the Count was a reasonable, enlightened man, who would retract his degeneracy theory if he were presented with overwhelming evidence against it. *Notes on the State of Virginia* was just one weapon in Jefferson's arsenal. Jefferson also wanted to present Buffon with tangible evidence—something the Count could touch. He tried with the skin of a panther, and then the bones of a hulking mastodon that had roamed America in the distant past, but Buffon didn't budge. Jefferson's most concerted effort in terms of hands-on evidence was to procure a very large, dead, stuffed American moose—antlers and all—to hand Buffon personally, in effect saying "see." This moose became a symbol for Jefferson—a symbol of the quashing of European arrogance in the form of degeneracy.

Jefferson went to extraordinary lengths to obtain this giant moose. Both while he was being chased from Monticello by the British in the early 1780s, and then later while he was in France drumming up support and money for the revolutionary cause in the mid-to-late 1780s, Jefferson spent an inordinate amount of time imploring his friends to send him a stuffed, very large moose. In the midst of correspondences with James Monroe, George Washington, John Adams, and Benjamin Franklin over urgent matters of state, Jefferson found the time to repeatedly write his colleagues—particularly those who liked to hunt—all but begging them to send him a moose that he could use to counter Buffon's ideas on degeneracy. Consider the following letter to former Revolutionary War general and ex-governor of New Hampshire, John Sullivan:

> The readiness with which you undertook to endeavor to get for me the skin, the skeleton and the horns of the moose . . . emboldens me to renew my application to you for those objects, which would be an acquisition here, *more precious than you can imagine*. Could I chuse the manner of preparing them, it should be to leave the hoof on, to leave the bones of the legs and of the thighs if possible in the skin, and to leave also the bones of the head in the skin with horns on, so that by sewing up the neck and belly of the skin, we should have the true form and size of the animal. However, I know they are too rare to be obtained so perfect; therefore I will pray you send me the skin, skeleton and horns just as you can get them, but most especially those of the moose. Address them to me, to the care of the American Consul of the port in France to which they come.[1]

The hunt for this moose, and the attempt to get it shipped to Jefferson, and then Buffon in Paris, is the stuff of movies. The plotline involved teams of twenty men hauling a giant dead moose through miles of snow and frozen forests, a carcass falling apart in transit, antlers that didn't quite belong to the body of the moose but could be "fixed on at pleasure," crates lost in transit, irresponsible shippers, and a despondent Jefferson thinking all hope of receiving this critical piece of evidence was lost. Eventually, though, the seven-foot-tall stuffed moose made it to Jefferson, and then to Buffon.

Because he saw so much on the line, Jefferson, as was his way, obsessed over providing every relevant fact to counter Buffon's anti-American theory of degeneracy; and his overall counterattack, including the moose, was powerful. At his 1826 funeral, one orator referred to Jefferson's efforts in this regard as the equivalent of leading a second American revolution. Thanks to Jefferson, refuting the theory of degeneracy was such a point of pride for early citizens of the United States that it was discussed in the opening pages of the country's first school textbooks.

Yet, despite Jefferson's passionate refutation, the theory of degeneracy far outlived Buffon and Jefferson; indeed, it seemed to have a life of its own. It continued to have scientific, economic, and political implications, but also began to work its way into literature and philosophy. On one side were those who continued to promulgate degeneracy—people such as the philosopher Immanuel Kant and the British poet John Keats, who described America as the single place where "great unerring Nature once seems wrong."[2] On the other side was a cadre that included Lord Byron, who spoke of America as "one great clime," Washington Irving, who mocked Buffon's theory in *The Sketch-Book of Geoffrey Crayon,* and Henry David Thoreau, who used his essay "Walking" as a platform "to set against Buffon's account of this part of the world and its productions."[3] This group saw America as a vast, almost unlimited land of resources, a place where nature shines on a world of healthy, hardworking people: and they labored (quite successfully) to make this idea part of our national identity. All of this can be traced back to the degeneracy argument between Buffon and Jefferson, and, to some extent to Jefferson's moose itself.

Eventually the degeneracy argument died; but it did not die an easy death . . .

CHAPTER ONE

"Dictatorial Powers
of the Botanical
Gentlemen of Europe"

To recant the degeneracy tale with the proper panache, we first need to understand who was involved in gathering information on American natural history in the eighteenth century, why they gathered this information, and how the study of natural history was conceptualized at the time.

In the Colonial era, as in every era, natural history information was, in part, passed along in what are known as travelers' tales.[1] These tales could be quite astonishing. In one, John Brickell, an Irish physician living in North Carolina, described how bear cubs were initially lumps of white flesh, "void of form," and only took on the shape of a bear as the result of their mother licking them, essentially molding a cub from a lump of formless flesh. For good measure, though, the same description noted that "the young cubs are a most delicious dish."[2]

Tales of North American beavers were equally incredible. A Louisiana engineer fascinated with dam building hid in the brush one night, and by the light of the moon watched beavers to see if he could pick up some tricks of the trade. Some of the beavers he observed made mortar from the mud, others lined up head-to-tail, loaded the newly made mortar on their tails, moved the mortar down this living assembly line, and applied it to the levee.[3]

Other travelers recorded sixty-three-pound turkeys as tall as a small man, and rattlesnakes that were twenty feet long and could, according to Jean-François Dumont, "bite off the leg of a man as clear as if it had been hewn down with an axe."[4] As late as 1770, the *Essex Gazette* newspaper published a letter reporting the discovery of a two-headed snake, with the head of a "yellow rattlesnake" on one end of its body and the head of a "black snake" at the other. The author, however, could not provide *Gazette* readers with as much detail as he had hoped because "the horrid form of the described creature urged the spectators to throw it precipitately into the river, which prevented a more critical examination."[5]

There were reports of lizards that, on their own volition, acted as reptilian guardians, protecting weary travelers: "If a person lies asleep, and any voracious beast, or the alligator . . . is approaching the place where you lie," wrote William Chetwood, "[the lizards] will crawl to you as fast as they can run, and with their forked tongues tickle you till you awake, that you may avoid by their timely notice the coming danger."[6] But such embellished folk stories of strange creatures made up only a small portion of the natural history data circulating in eighteenth-century America. People of that day were interested in learning about natural history because it could have the immediate and utilitarian impact of improving everyday life.[7] Travelers' tall tales were clearly useless in this regard, and were more entertaining than functional.

★ ★ ★

Natural history information was valuable to early Americans for many reasons. Learning about the habits of prey such as rabbits, squirrels, deer, and partridge could make the difference between feast and famine. Information on which species of snakes were venomous might be a matter of life and death. The function of natural history was especially evident in the area of botany, where new information could lead to healthier crops, better tasting foods, possible cures for illness, or even a combination of these effects. Paul Dudley, in a 1720 paper published in the *Philosophical Transactions of the Royal Society of London*, described the process of removing sap from maple trees in New England, noting that the sugar from this sap had medicinal value that was greater than that associated with sugar from the West Indies.[8]

Many of the medicinal applications of botanical natural history were gathered together by Philadelphia's Benjamin Barton in his 1788 textbook, *Collections for an Essay toward a Materia Medica of the United States*.[9] Of course, some of these practical applications of natural history were misguided or plain wrong, and some were even unintentionally dangerous.

For the most part, America's knowledge of plants and animals was gathered by men who were natural historians by avocation, but not by profession—men such as Dudley, Barton, Henry Muhlenberg, Manasseh Cutler, James Logan, and Cadwallader Colden.[10] They almost always lacked royal patrons or other moneyed contacts, and they often held other jobs—teacher, minister, businessman, politician, and very often physician—gathering information about the natural history of America when time permitted (and sometimes when it didn't).[11] They were passionate about the pursuit of natural history, collecting when they traveled, imploring their colleagues, both at home and abroad, to send them any information they may have acquired, and, on the botanical end, occasionally keeping their own herbaria.

At the start, American natural historians were at a severe disadvantage, lacking the royal gardens, the zoological collections, and the ancient university system that were available to their European counterparts. What's more, the ultimate accolade for a natural historian—membership in the prestigious Royal Society of London—was also unavailable to these men. While some American natural historians were elected to the Royal Society, policy dictated that the names of Colonial members were not printed on the list of "Fellows of the Royal Society."[12]

Many natural historians in the colonies (and early states) were driven, in part, to see that the authoritative works on American natural history would one day be seen as having been penned by Americans. They were tired of their European counterparts looking down on their efforts: what was needed was an end to what one American naturalist, with the wonderful name of Alexander Garden, called "the dictatorial powers" of the "botanical gentlemen of Europe."[13] Along these lines, Henry Muhlenberg implored his friend Manasseh Cutler to "let each one of our American botanists do something and soon the riches of America will be known."[14] But at the beginning of the nineteenth century, the state of natural history in America, at least in Mulhenberg's eyes, was still far from where it needed to be: "The study of natural history in this country is in its infancy. . . . We have no cabinets of natural history in America, excepting one in Philadelphia and another in Boston. These consist of small collections without any systematic arrangement. They are kept merely for the purpose of getting money by showing them to common people, and consist primarily of exotics."[15] This may not have been a completely fair representation, at least with respect to Charles Peale's Philadelphia natural history museum—which contained thousands of animal samples and was a tourist attraction—but it was an accurate representation of the overall state of affairs.

Eventually, and in particular after the War of 1812, the study of American natural history would blossom under the eyes of such men as Asa Gray, John James Audubon, Thomas Nuttall, and William Maclure, many of whom followed in the footsteps of the great ornithologist Alexander

Wilson. These men and others associated with the Philadelphia Academy of Natural Sciences would produce volume after volume of beautiful tomes on the natural history of the United States.[16] But, before the nineteenth century, the systematic study of natural history was primarily a European affair and was associated with two of the continent's most famous natural historians: Buffon and Linnaeus.

The fact that most of the work on eighteenth-century natural history was coming out of Europe tends to overshadow an important point about the way that Americans conceptualized the study of nature. From the few rich enough to own books on this subject to those dependent on knowledge of it for survival, when Americans thought about nature, when they talked about natural history, they did so with supreme confidence that the life they saw around them—the animals and plants—were designed by God, and just as importantly, were a manifestation of God's perfection.

To understand natural history was as close as humans could come to understanding something about the divine. More than a century before Darwin would posit a purely naturalistic theory for understanding life on earth,[17] the view that natural history was a reflection of God's wonder was not just one way to think about the world; it was the predominant way.

This idea of nature as a reflection of God's power has a long history.[18] When Plato contemplated both the organic and inorganic worlds around him, he could not accept that they were the result of chance, for they displayed what appeared to be order and complexity; the world appeared to be designed. This implied a designer—Plato used the term *demiurge*.[19] To understand the world, then, was to understand something about its designer.[20] This religious/philosophical position is often called the argument from design.

Over the next few centuries, the argument from design was adopted by major thinkers of the day, such as the Roman statesman Cicero (106–43 BC), who asked how anyone could doubt that the world, and all the species that occupied it, were the result of "conscious intelligence,"[21] to Galen (circa AD 129–200), one of the founders of modern medical thinking, who wrote that animals were "fully equipped with the best possible bodies" and was impressed with "the skill employed in the construction of each."[22]

The argument from design became linked with a Christian god in the work of Saint Augustine (AD 354–430). In book 11, chapter 4, of his *City of God*, Augustine proclaimed "The world itself, by the perfect order of its changes and motions, and by the great beauty of all things visible, proclaims by a kind of silent testimony of its own both that it had been created, and also that it could not have been made other than by a God ineffable and invisible in greatness and invisible in beauty." Similar sorts of proclamations can be found in the writings of Saint Thomas Aquinas (AD 1224–1274). The study of natural history—of the wonders of animals and plants—was now linked with a better understanding of divine inspiration.

Though the argument from design would eventually draw some detractors—men like David Hume[23]—its allure was powerful. Even Francis Bacon, who in general advocated a strong separation between science and religion, was a firm believer that natural history was a window to God. Bacon argued that we have two sources to comprehend the divine plan, "The Scriptures, revealing the Will of God; and then the creatures expressing His Power; whereof the latter is a key unto the former."[24]

The argument from design that would eventually set sail for America came to maturity in Englishman John Ray's 1691 treatise entitled *The Wisdom of God Manifested in the Works of the Creation*. Though other Brits in Ray's time described how natural history was a reflection of God's plan,[25] *The Wisdom of God* stood out for two reasons. First, Ray was not only an ordained minister, but also one of England's leading naturalists. Just as important, though, Ray wrote a book that was meant for public consumption. His message to the reader was clear:

> to illustrate some of His principal attributes; namely, His infinite power and wisdom. The vast multitude of creatures, and those not only small, but immensely great . . . are effects and proofs of His Almighty power. . . . The admirable contrivance of all and each of them, the adapting all the Parts of animals to their several Uses, the provision that is made for their sustenance, which is often taken Notice of in Scripture. . . . And, Lastly, their mutual subserviency to each other, and unanimous conspiring to promote and carry on the public good, are evident demonstrations of His sovereign wisdom.[26]

For John Ray, the evidence that natural history reflected God's greatness could be seen everywhere. How else could one explain the vast number of different life forms? What other possible explanation could there be for the beautiful architecture of a bee's nest, the craftsman-like skill of the beaver's dam, the uncanny ability of the chameleon to match its environment? None of these incredible observations made any sense without the hand of God. "In Wisdom hast thou made them all," Ray wrote; " . . . the Works of God, they are all very wisely contrived and adapted to Ends both particular and general."[27]

In early Colonial America, this idea of natural history as a window into the divine was championed by the Reverend Cotton Mather, the most famous clergyman of his day. Mather had his hands in virtually every component of life in New England: from the Salem witch trials of the 1690s to the 1721 smallpox epidemic, where his championing of inoculation as a means of protecting people was viciously attacked by many. A Puritan Congregational leader, Cotton Mather was a prolific author, writing some four hundred works, including *Essays to Do Good* (1710) and a seven-volume magnum opus on the history of religion in New England entitled *Magnalia Christi Americana* (1702).[28]

All life reflected God's glory in Mather's writing. In his 1693 book, *Winter Meditations*, he summarized a lifelong view that the almighty was "in all the creatures . . . directing all their motions . . . to God himself as the last end all."[29] But it was in his 1721 book, *The Christian Philosopher: A Collection of the Best Discoveries in Nature with Religious Improvement*, where Mather most powerfully linked an understanding of natural history with religion.[30]

Mather was a not a natural historian himself; he rarely went out into nature to collect and describe species. *The Christian Philosopher*, then, was not a religious interpretation of Mather's personal interactions with nature. Instead, Mather relied heavily on the most up-to-date information he had read or heard, and interpreted it in the context of what it revealed about the divine architect of the world. And he made no apologies for using the findings of others, such as John Ray.[31] Indeed, Mather believed that he was spreading the name of God by doing just that, for "it appears also but a piece of justice, that the names of those whom the great God has distinguished, by employing them to make those discoveries, which are here collected, should live and shine in every such collection."[32] His reliance on the work of others certainly did not limit the scope of the topics Mather would cover in *The Christian Philosopher,* whose chapters included everything from "Of Light" to "Of Insects," "Of Reptiles, "Of Fishes," "Of Birds," "Of Four-footed Animals," and "Of Man."

Mather had both a goal and a strategy in *The Christian Philosopher*. The goal was to demonstrate that "the whole world is indeed a temple of God, built and fitted by that Almighty architect."[33] And, of course, the whole world included all those species with which humans share the planet. The strategy he would adopt to prove this was to assure his readers that the findings of natural historians—men who were often called "natural philosophers" in his time—were not a threat to religion, but, rather, when properly construed, evidence of God's perfection. "The essays now before us," Mather wrote in the introduction to *The Christian Philosopher*, "will demonstrate that philosophy is no enemy, but a very great incentive to religion."[34] Mather saw scripture as just one of the twofold books of

1.1. The Reverend Cotton Mather, author of *The Christian Philosopher: A Collection of the Best Discoveries in Nature with Religious Improvement*. Portrait by Peter Pelham.

God. The second, "the book of the creatures," was what he would focus on in *The Christian Philosopher*.

Cotton Mather was genuinely amazed by the world around him. This was evident in every chapter of *The Christian Philosopher*, including the chapter concerning birds, which read more like a tribute to all things avian. People had long been awestruck by the ability of birds to fly, but for Mather, there was so much more to flight than met the eye, and all that he learned pointed to evidence of divine handiwork. "How commodiously their wings are placed!" Mather told his readers, and no doubt his parishioners as well: "They that fly much, or have most occasion for their wings, have them in the very best part imaginable, to balance their body in the air, and give them a swift progression. . . . The incomparable curiosity of every feather!"[35]

Flight was not the only avian attribute that would lead to such prose: "Birds that climb, as the wood-pecker kind, how fitted for the purpose!"[36] The very way that birds packaged their offspring into nicely rounded, protected eggshells also evoked unbridled praise: "That birds must lay eggs, is a sensible argument of a divine Providence, designing to preserve them, and secure them, that there might be a greater plenty of them," begins Mather. Why? Because had they given birth to live young, rather than laid eggs, they might have "brought forth a great number at a time," and consequently, "the burden of their womb would have rendered them so heavy, their wings could not well have served them."[37]

Even the bloodier side of bird life was evidence of God's design and intention:

> Shall we stop a moment, and consider how useful the carnivorous birds of prey become, even in prosecuting their voracious inclinations? If the number of lesser birds were not by their means lessened into such a proportion, those lesser birds would overstock their feeding; and then also, should those lesser birds, which are so numerous, die of age, they would leave their carcasses to rot upon the ground, corrupt the air, and become insupportable.[38]

To Mather, every attribute of bird life—the flying, the laying of the eggs, and even the killing that needed to be done by carnivorous species— was evidence of divine architecture. "Such essays as these," he wrote, "to observe, and proclaim, and publish the praises of the glorious God, will be desirable and acceptable to all that have a right spirit in them; the rest, who are blinded, are fools and as little to be regarded as a monkey flourishing a broomstick!"[39] Mather pleaded with skeptics who might read *The Christian Philosopher*: "Blind philosopher, canst thou see no God in all this?"[40] Eighty years later, another reverend—William Paley—would make a similar case for the argument from design, but Mather and Ray were among the first to do this, and package it in a way to reach the general population.[41]

That nature represented the manifestation of God's design was a notion that eighteenth-century American naturalists tended to cling to tightly.[42] And though the vast majority of the American population may not have read Plato, Augustine, or Bacon, many would have known of Cotton Mather. And even if they had never heard of Mather, they would have been taught the design argument at home, and in sermons at church. It was the way most people thought about nature, and hence natural history.

It was not only those who held Mather's strict views on such matters as the truth of divine revelation who were drawn to the design argument; Deists—who believed in one God, but rejected divine revelation and the notion that God intervened in the contemporary world—acknowledged it as well. The argument from design was well in step with Deist views. Deists, too, saw the result of God's handiwork all around them. A case in point was Thomas Jefferson, who, though raised an Anglican, was strongly influenced by Deist works from John Locke and others.[43] Jefferson usually made his personal religious beliefs a very private affair, but he made an exception late in life when he outlined his Deist tendencies in a letter to John Adams: "I hold (without appeal to revelation) that when we take a view of the universe . . . it is impossible for the human mind not to perceive and feel a conviction of design, consummate skill and indefinite power"—a designer that Jefferson called the "fabricator of all things . . . their preserver and their regulator."[44]

The argument-from-design mindset was one of the reasons that Buffon's theory of degeneracy was viewed as so vile by Americans. If nature reflected divine architecture, the claim that all species in America were inferior to life elsewhere had to be an affront. And, Buffon's degeneracy argument challenged not only a fundamental belief about God and nature, but it had political, economic, and social ramifications—none of which were good for the fledgling United States of America.

CHAPTER TWO

The Count's Degenerate America

The animals file past him one after another as though they were just coming forth from Noah's ark. One by one the naturalist [Buffon] looks them
over and, and each in turn is denied American citizenship.[1]

Such is the picture painted of the greatest naturalist of his time, the
French Count Buffon, as he composed his massive, thirty-six-volume
Natural History: General and Particular (*Histoire Naturelle*).[2] The lion, king
of the beasts, was to be found in the Old World; nothing resembling its
magnificence existed in the New World. The giant elephant called the
Old World its home. The best the Americas could produce was a tapir,
who was but a sixth the size of the great elephant. In his survey of life in
the two worlds, Buffon went on and on with such comparisons, and the
New World always came up deficient. For Buffon, all animals in the New
World were degenerate—smaller, and less robust—than those found in
the Old World. All that the Americas could lay claim to was being stock
full of insects, snakes, and frogs. But it was hardly the fault of these New
World creatures that they existed in degenerate form.

In Buffon's mind, the fault lay with the Indians who had long ruled the
land. They were cold, lazy, without feeling, stupid, and lacking in sexual
drive. They had failed to conquer nature, and so the land remained wet
and cold. And for Buffon, wet and cold environments led to degenerate
life forms, leaving him free to make sweeping claims about the superior
life forms to be found in the Old World—which, of course, included his
own beloved France.

2.1. Count Buffon, author of *Histoire Naturelle*. The Buffon oil painting is by François-Hubert Drouais.

The birthplace of these arguments about New World degeneracy—Buffon's *Natural History*—is a massive encyclopedia that was published over the course of nearly four decades (1749–1788). The title understates Buffon's grand vision for this work. It was more than a catalog of information about species from across the planet; for unlike anything that came before it, *Natural History* presented sweeping characterizations of all the species it described.

More than mere summaries of each species, Buffon's concept of natural history was meant to capture each organism in the context of the environment in which it lived. Yet he was not satisfied with even this expanded treatment of life on earth; that was only a slice of what *Natural History* set out to cover. In the opening volumes of his ambitious undertaking, Buffon criticized those who claimed that there was scientific evidence for Noah's flood. He laid out a naturalistic hypothesis about the creation of the earth in which a comet struck the sun, tore away bits of it, and formed the planets. He also rejected contemporary ideas that all life already existed "preformed" in eggs and sperm, arguing that taking such a stance was "not only to admit that one does not know how it happens, [but to] abandon the will to think about it."[3] In its place, Buffon viewed reproduction as a special case of growth ensuing from sperm and egg uniting. And finally, Buffon used the first volumes of *Natural History* to study humans, by using the same natural history methods he employed to study other species.

By delving into such topics as Noah's flood, the birth of planets, and the history of humans—topics that were generally reserved for the church—Buffon's *Natural History* raised questions about his own belief system. Historians of science still debate whether or not Buffon was an atheist.[4] He was certainly accused by many—most famously by the Deputies and Syndic of the Faculty of Theology of Paris—of contradicting fundamental precepts of Christianity, particularly with respect to his materialist views on the origins of the universe.[5] But Buffon himself denied contradicting the teachings of the church.[6] Though he was uncomfortable with the idea of miracles, and he denied the direct hand of God in creating the earth, he also spoke of man's superiority over the beasts by light of reason and possession of a soul.[7] What is clear is that Buffon's views on whether God existed, and what role he might play in explaining the universe, changed numerous times throughout the course of his life, and that *Natural History* (and Buffon's other works) were secular works. *Natural History* was meant to describe the world without reference to supernatural causes, and this was the way it was received by both Buffon's supporters and his critics.[8]

It was in this all-encompassing work on nature that Buffon set out his theory about degenerate animals and humans in America.

★ ★ ★

Georges-Louis Leclerc, who would one day attain the rank of Count of Buffon,[9] was born on September 7, 1707, in the village of Montbard, in the Burgundy section of France.[10] He was the eldest son of Benjamin-François Leclerc and Anne-Christine Marlin. His father was a respected local lawyer, but it was not until his mother's wealthy uncle, Georges Blaisot, died and passed his fortune on to the Leclercs, that the family began to amass real power in the region. Uncle Georges' money was used to buy the nearby village of Buffon, as well as to purchase the rank of counselor to the Burgundy Parliament. Though they maintained ownership of their estate there, with their newly established wealth and power, the Leclercs left Montbard for Dijon when Buffon was thirteen.

Not much is known about Georges-Louis's early days, and his biographer, Jacques Roger, notes that Buffon himself "seems to have wanted to repress all memories of his childhood."[11] Once the family moved to Dijon, Buffon attended the College de Godrans, a Jesuit institution, which provided him with a solid background in both mathematics and the physical sciences. His lifelong passion for mathematics was conceived there, but Buffon's classmates at the College de Godrans remembered him more for his athletic abilities than his intellectual prowess.

In 1723, Buffon, likely at the insistence of his father, entered the law college in Dijon, and formed his first long-lasting friendships, including one with a lawyer named Jean Bouhier, who was also president of the local parliament. Bouhier's library was very large, and it was there, through his readings at his friend's house, that Buffon became immersed in the works of philosophers and mathematicians. In 1726, at the age of nineteen, he graduated law school and, immediately after, decided to abandon law as a profession. The mathematics he had learned aroused an intense interest in the subject in young Georges-Louis, and he saw this, rather than the law, as the path to his future. Because of the standing of his family, Buffon had the leisure to pursue his interests, and he soon began contacting some of the leading mathematicians in Europe,[12] eventually moving to Angers to study mathematics and immerse himself in the work of Newton. His stay in Angers was cut short in 1728, apparently because of romantic difficulties that forced him to return to Dijon.

In the late 1720s and early 1730s, Buffon's passion for mathematics grew. One of his early, and long lasting, interests centered on probability theory and, more specifically, on games of chance. Part of this was due to a deep fascination with the subject itself, and part was probably related to his passion for money and maintaining his wealth. Buffon was not a big gambler himself, but the more he could understand about money, the better, as he was involved in a nasty lawsuit (dealing with his father's remarriage to a very young bride) that threatened his inheritance during the same period that he was working on problems in probability. With time, Buffon made a name for himself based on his work on games of chance and related matters in probability, and his skills in these areas would serve him well when he began writing *Natural History* decades later.

Buffon moved to Paris in 1732, and while there, lived at the home of Gilles-François Boulduc. Boulduc was apothecary to the king, professor of chemistry at the Royal Garden, and most importantly from Buffon's perspective, a member of the Royal Academy of Sciences. Always ambitious, Buffon saw his relationship with Boulduc as a means of gaining entrance into the Royal Academy. He felt entitled to such consideration, and fortunately for him, entrance into the academy was possible at a very young age. The lowest rank in the Royal Academy—that of "assistant"— was at times filled by twenty-year-old men, and the higher ranks were often filled by men no more than twenty-five years old.

Based on his association with Boulduc, and the allies he had already made in mathematical and philosophical circles, by January 1733 Buffon boasted to friends that his admission to the Royal Academy was imminent. Before that, however, Buffon would need to present a paper to the academy and be judged in person. He chose to present on a game of chance called *Franc-carreau*, in which players bet on where a coin will land on a floor full of tiles.[13] Buffon's paper was well received; he was invited back a few months later, and then was chosen to fill an assistant position that had opened in the academy's section in "mechanics."[14] Soon after that Buffon began a yearly migration pattern that would last the rest of his life: the fall and winter were spent at the academy in Paris, and the spring and summer at his family estate back in Montbard.

Of Montbard and Paris, Buffon's preference was clear; "Paris is hell," he wrote, "and I have never seen it so full and so stuffed. I, unfortunately, do not have a taste for sticky involvements; at any moment, there are many endlessly going on. I would rather spend my time making water flow and planting hops than to waste it here with useless errands and more uselessly paying court."[15] He endured Paris for the academy, and equally importantly, because his ambition had convinced him that to wield power and be economically successful, he needed to spend time there.

Buffon tolerated Paris, but he longed for his family estate. And it was there, in Montbard, that his interest in natural history was kindled by a rather surprising series of events. In the early 1730s, the French navy was investing in research to increase the strength and longevity of wood to make its fleet even more daunting, and they turned to the Royal Academy of Sciences for advice. Because Buffon's family estate at Montbard was surrounded by beautiful forests, he was approached about researching the properties of wood and agreed to undertake some experiments on the matter. Although he sold his forests for a hefty profit just a year later, Buffon was named "director" of these lands for life,[16] and he conducted experiments in the forests of Montbard for the next forty years.

While much of the work done in the forests around Montbard was of the practical nature dictated by the navy, Buffon developed a genuine interest in botany as he performed experiments. He saw a need not only to observe and record, but to gather a deep knowledge of the system he was studying. Buffon was not interested in individual trees, but rather in the way that forests, as entities, operate. His new love of botany helped shape a view of nature that he would develop more fully in *Natural History*.

Never one to attack a subject halfheartedly—he typically worked fourteen hours a day, seven days a week—Buffon immersed himself in botany, and his passion quickly led him to translate Stephen Hales's book

Vegetable Statistics from English to French. He admired Hales's book not just for the botany, but because of Hales's approach to science, which involved "keen, reasoned and sustained experiments. . . . Nature is forced to show us her secrets: all other methods have never succeeded. . . . This is the method . . . of the great Newton. . . . It is that which the Academy of Sciences has made a law to adopt. . . . In short it is the path that has led great men for all of time, and which still leads them today."[17] And this was the path that Buffon, too, would follow.

In 1739 Buffon transferred from the mechanics section of the academy to the botany section and was promoted from assistant member to associate member. At about the same time, based on his contributions to both mathematics and botany, he was elected to the Royal Society of London. But before Buffon could truly savor these accolades, a surprising turn of events, which would make the Royal Academy and Royal Society memberships pale in comparison, would land him in a position he could hardly have imagined months before.

On July 17, 1739, Charles de Cisternay du Fay, curator[18] of the Royal Botanical Gardens, died of smallpox. Curator of the Royal Botanical Gardens was one of the most prized positions in France, and almost immediately, Buffon began campaigning for the job. He implored a colleague[19] that "even though I would have more reasons to claim it than another, I would not dare ask for it. . . . I will pray my friends to speak for me. . . . One might realize that the intendancy of the Royal Botanical Gardens needs a young, active man who can brave the sun, who knows the plants and the way to multiply them, who is somewhat knowledgeable in all areas that are asked of him. . . . It appears to be thus that I am what they are looking for."[20] Buffon's friends approached the king on his behalf, who in turn appointed Buffon—at a regal salary[21]—to the position at the Royal Botanical Gardens on July 25, 1739.

As soon as he assumed the position of curator, Buffon began cataloging the specimens not only of the Botanical Gardens, but of all the king's "cabinets," including the Cabinet of Natural History. This marked the starting point for a massive encyclopedia of natural history that he planned to write. Ten years would pass before the first volumes of that *Natural History: General and Particular* would appear; and though Buffon didn't know it in 1739, this monumental work would engulf almost every moment of his waking hours for the next forty-seven years.[22]

Published by the Royal Press of France between 1749 and 1788, the thirty-six volumes of *Natural History: General and Particular,* were designed to present the reader with what Buffon called "the exact description and the true history of each thing."[23] The work was an immediate success:

the first printing of the early volumes sold out in six weeks, followed by a second printing, and then a duodecimo edition similar to a modern paperback. *Natural History* was quickly translated into English, Dutch, and German. It was *the* book to have in the salons of Paris, and was widely regarded as one the eighteenth century's best sellers—a startling accomplishment, since all the natural history works published before Buffon were dry, fairly technical monographs that were anything but best sellers.[24]

As new volumes of *Natural History* were published, Buffon's arguments would be picked up in European newspapers and journals, such as the *Journal de Trévous*, which began its review, "This great work . . ." and went on to describe "the ingenious capacious and pleasant fashion in which the details are set forth."[25] Though there were European skeptics, Buffon's work was generally admired on the continent. *Natural History's* allure was not restricted to Europe, however. Dozens of American newspapers, including the *Massachusetts Gazette,* the *Massachusetts Centinel,* the *Pennsylvania Packet,* the *Maryland Herald,* the *New Jersey Journal,* and the *State Gazette of South Carolina*, also wrote about Buffon's encyclopedia, often embedding their reporting within critiques of the degeneracy argument.[26] The tone of these articles was that *Natural History* was, in general, a masterful work—except for the sections on degeneracy, which were described in the *State Gazette of South Carolina* as a "humiliating picture indeed."[27]

Buffon's popularity skyrocketed. He was depicted in paintings as having conversations with Mother Nature and had a cult equal to that surrounding Voltaire and Rousseau—the king ordered his statue to stand in front of the Royal Gardens, where it remains to this day. So enormously popular was the author of *Natural History* that when he died, the *Mercure* newspaper wrote of a funeral march where "twenty thousand spectators waited for this sad procession, in the streets, in the windows, and almost on the rooftops, with that curiosity that the people reserve for princes."[28]

Natural History is meant to be read in blocks. Volumes 1–3 address the formation of the planets, including the earth, some general thoughts on animal biology, and a discussion of man and his place in nature. Volumes 4–15 are devoted to four-legged creatures (quadrupeds); volumes 16–24 examine birds; volumes 25–31 are "supplements"; and the last volumes focus on minerals. It was the twelve-volume section of *Natural History* devoted to quadrupeds (henceforth denoted as *Quadrupeds*) in which Buffon would present his arguments on degeneracy in America. When Buffon spoke of "America," he was referring to both North and South America, but our focus is on the North, especially the United States colonies.

In *Quadrupeds,* as well as in the rest of *Natural History*, Buffon rejected the commonly held notion that the job of the natural historian was merely

2.2. Statue of Buffon at the Royal Garden (by Jean Carlus).

to describe and classify, as his rival Linnaeus had done. Buffon took the "history" part of natural history very seriously. This meant tracking organisms from birth to death, and understanding what they did during the whole of their lives—Buffon set out to give "the true history of each thing," and "not the history of the individual, but that of the entire species." He took on the monumental task of describing "their generation, the duration of pregnancy, birth, the number of young, the care of the fathers and mothers, their type of rearing, their instinct, the place of habituation, their diet, the way they obtain it, their habits, their tricks, their way of hunting, and then the services they can render us, and all the uses and commodities we can obtain from them."[29]

To understand how nature, writ large, operated, Buffon saw the job of the naturalist as "glancing behind and ahead, to try and catch a glimpse of what she [Nature] could have been before, and what she may become in the future."[30] This, he recognized, was no easy task, which is why no one else had ever tried producing an encyclopedia on the scale of that of *Natural History*. To accomplish such a task, Buffon argued, required a set of traits that are rarely found in one man, namely, "the broad view of a fervent genius that embraces everything in a single glance, and the attention to small details of a laborious instinct that looks at only one detail at a time."[31] He was certain that he was that rare fervent genius.

In *Quadrupeds,* Buffon, in collaboration with Louis-Jean-Marie Daubenton,[32] who wrote the anatomical and morphological sections in these volumes, presented the complete natural history of species after species: more then six thousand pages worth. Buffon often began by presenting the reader with the name of the species in question in as many languages as the literature provided; for instance, he presented the common name of the deer in Greek, Latin, Spanish, Portuguese, English, German, Polish, Swedish, Danish, Dutch, Russian, Turkish, Persian, and Arabic, as well as other languages. This was followed by as complete a synopsis of behavior, reproduction, geography, anatomy, and morphology that Buffon could uncover. And embedded in all this were wonderful drawings that, to this day, still fetch huge sums from art collectors.

Spending years in far-off lands studying exotic species was not the modus operandi of most naturalists in the eighteenth century. In Buffon's era, natural historians did not usually spend time out in nature themselves making observations. Indeed, Buffon had the reputation of *never* having made observations on the species he described in *Natural History,* and he did little to squelch this rumor.[33] If he was not actually out in nature gathering data, where then did Buffon come up with thousands upon thousands of pages of information on quadrupeds? The prime source for much of the anatomical and morphological descriptions in *Natural History* was the Royal Cabinet itself. In addition to what was there when Buffon took over in 1739, he had considerable sway with royal ministers, and was able to procure significant funds to add specimens to the cabinet. In some instances, this amounted to adding huge intact collections.[34]

Buffon was also able to gather data directly on some species, albeit in very unnatural settings. One such setting was his family estate, where he cordoned off one area and attempted to create a "semi-wild" environment that he stocked with foxes, hedgehogs, cats, chickens, dogs, badgers, and a monkey named Jocko. Though Buffon was able to gather some data at Montbard, most of the time his collection of animals went about chasing each other, burning themselves near fires, and begging their keepers for food. Buffon gathered a bit more reliable data at the Royal Menagerie, where he verified what he had heard about zebras and elephants by observing them in person.

The Count relied heavily on the accounts of travelers for obtaining information to include in *Natural History*. It was common practice for those traveling on business ventures, or as missionaries, or on some government/military mission (French or otherwise), to take notes on the animals they encountered, particularly exotic species. As head of the Royal Gardens, Buffon had access to many such accounts, and he solicited these sorts of reports from a variety of individuals. Buffon would then compare the notes of travelers and distil them down to some general description, eliminating what appeared (to him) to be outrageous and unsubstantiated claims along the way. This vetting process was far from perfect, especially because travelers would give different names to the same species, as well as embellish their tales for the sake of potential readers. At the very least, though, Buffon was clear whether the descriptions that he provided in *Natural History* were based on his own work or that of others.

Natural History was anything but a dry work. Instead, it was designed with the educated and semiaffluent reader in mind, and that is who it reached. This audience of readers would become the mainstay of Victorian natural history authors a century later, but it was first tapped into by Buffon. Unlike the tedious reading associated with most of the other scientific books of Buffon's day, it became chic to read *Natural History*.

Buffon's goal was to make his readers *think* about nature and natural history, and not bog them down in ways that other enlightened philosophers seemed to savor. To do this, he would write simply, but eloquently, drawing in the reader along the way. For example, in order to get a general flavor for how to think about the natural world, Buffon would ask the reader of *Natural History* to "imagine a man who has actually forgotten everything or who wakes up innocent of everything around him; place that man in a countryside where the animals, the birds, the plants, the stones present themselves successively to his eyes. . . . Soon he will form a general idea of animated matter . . . and naturally he will arrive at the first great division, animal, vegetable and mineral."[35] If readers wanted much more in the way of details, they could find it, especially in the sections written by Daubenton, but they could also choose to skip the minutiae and walk away with a general sense of what Buffon had to say about a topic.

As more and more volumes of *Natural History* were released, Buffon's rhetorical skills became legendary.[36] His description of the wolf, for example, has an almost romantic and mythic component to it:

> The Wolf is one of those animals whose appetite for animal food is very strong. Nature has furnished him with various means for satisfying this appetite, and yet though she has bestowed on him strength, cunning, agility, and all the necessary requisites for discovering, pursuing, seizing, and devouring his prey, he not unfrequently dies with hunger; for man having become his declared enemy, and put a price upon his head, he is obliged to take refuge in the forests, where the few wild animals he can meet with escape him. . . . He is naturally dull and cowardly, but becomes ingenious from want, and courageous from necessity. . . . He braves

danger. . . . He conceals himself during the day in his den, and only ventures out at night, when he traverses the country, searches round the cottages, kills such animals as have been left without, scratches up the earth from under the barn-doors, enters with a barbarous ferocity, and destroys every living thing within. . . . The young wolf may be tamed, but never has any attachment. Nature in him is stronger than education; he resumes, with age, his ferocious disposition, and returns as soon as he can to his savage state.[37]

Buffon understood the value of his eloquent rhetorical style. When inducted into the Académie française on August 25, 1753, he presented a seminar which contained his famous line "Style is the man himself" ("Le style est l'homme même"). But Buffon's rhetorical skills came back to haunt him. It seemed to his colleagues that Buffon's writing abilities were a bit *too* sharp, and that his popularity and that of *Natural History* were more a matter of style than substance—something that most savants did not appreciate.

Georges Cuvier, naturalist, anatomist and paleontologist who along with Buffon was considered one of France's greatest Enlightenment thinkers of the seventeenth/eighteenth century, called Buffon's writings "pompous."[38] Others were more discreet and couched their comments in backhanded compliments. Charles Bonnet, a well-respected Swiss natural historian and philosopher of the day, wrote that Buffon was "a sublime and bold genius," but that he was "too carried away by the spirit of the system," and that *Natural History* was close to a "philosophical novel."[39] Rousseau was impressed with Buffon the writer, but not Buffon the thinker: "His writing will instruct me and please me my entire life. I believe him to have equals among his contemporaries in his quality as a thinker and philosopher, but for his quality as a writer, I know none [equal] to him."[40] These sorts of critiques carried into the twentieth century, with historian of science Arthur Lovejoy noting that "Buffon was little careful of consistency—and extremely careful of rhetorical effects."[41]

This distinction between style and substance in Buffon's *Natural History* would become important in the argument over American degeneracy. Thomas Jefferson, in his reply to Buffon's claims of American inferiority, was quick to point out that "there has been more eloquence than sound reasoning displayed in support of this theory," and that the Count's readers had "been seduced by a glowing pen."[42]

The dozen volumes of *Quadrupeds* were written in the form of a species-by-species description of the natural world. Interspersed within these volumes are four more general treatises, and it is in these that Buffon lays out his grand theory of animal and human degeneration in the

New World. Volume 9 contains three of these treatises—"Dissertation on Animals Peculiar to the Old World," "Dissertation on Animals Peculiar to the New World," and "Dissertation on Animals Common to Both Continents," while volume 14 has one: "Treatise of Degeneration of Animals."

In applying his degeneracy theory to animals, Buffon made four related claims. First, he argued that animals found in both the New World (read America) and the Old World were smaller and feebler—degenerate—in the New World. Second, animals found only in the New World were degenerate compared to those found only in the Old World. Third, there were fewer species in the New World, and those which existed only there tended to be "cold-blooded" insects and reptiles; and fourth, domesticating animals caused degeneration, especially when that domestication occurred in the New World. As he developed this litany of New World animal degeneration, Buffon extended his claims to humans, or, more specifically, to one type of human—the indigenous American Indian.

Buffon was relentless when it came to degeneration in New World animals: he saw no exceptions to the law of nature he had discovered.[43] While the theory of degeneracy would be indelibly linked with his name, Buffon was not the first to suggest that there was something inherently inferior about the New World and the life it contained.[44] This idea can be traced back as far as Hippocrates and Aristotle (who spoke of people in "cold countries" being "impulsive),[45] and Buffon's readers would have been aware of this history. And by the sixteenth century, the New World had already been linked with lowly insects and reptiles.[46] More generally, as Antonello Gerbi argued in his *The Dispute of the New World,* before Buffon, the inferior nature of New World animals had "been expressed as curious revelations of distant lands in the descriptions of the early travelers and naturalists in the New World, or as polemic paradoxes and fables in the reports of the missionaries, in . . . utopias and myths."[47] And pre-Buffonian ideas on the degeneration of New World humans were along the same lines. When Queen Isabella learned that the climate of the New World was so humid that the trees could not grow their roots deep into the soil, she feared that "this land, where the trees are not firmly rooted, must produce men of little truthfulness and less constancy."[48] The French philosopher Baron Montesquieu (1689–1755), too, would argue for the dramatic effect of climate on both animals and humans.[49]

Buffon's New World degeneracy argument was fundamentally different from the ideas on New World inferiority that preceded it. Early ideas were vague, amorphous, and undeveloped, and rarely if ever posited a cause for this presumed inferiority. Buffon provided legions of specific examples, and proposed a set of clear causes for degeneracy. More importantly, though, what fundamentally separated Buffon's arguments for New World degeneracy from those that preceded it were author and context. As he published *Natural History,* Buffon became one of the most widely respected natural historians in Europe, which is to say that the idea that the Old World was superior to the New World—and for our story, particularly America—was now emanating from one of the shining stars of the Enlightenment. What's more, the degeneracy claim was couched in *Natural History,* itself a massive review of the entire history of life. It appeared then not as some wild idea, but one that naturally flowed from the most thorough study of life ever undertaken.

Buffon began his in-depth analysis of New World degeneration in "Dissertation on Animals Peculiar to the Old World." The tone of Old World superiority that was to run throughout all the sections relating to degeneracy was set immediately, as the reader was told at once: "Elephants belong to the Old World. . . . It is unknown in America, nor is there any animal there that can be compared to it in size or figure. The same remark applies to the Rhinoceros; . . . we have seen that the lion exists not in America, . . . and we shall now find that the tiger and panther belong also to the old continent." Comparison was a key motif for Buffon; after describing the tiger, he quickly contrasted it with the "animals of prey belonging to the new continent, as the largest of them scarcely ever exceed the size of our mastiffs."[50]

Buffon chose this opening list of creatures carefully: these large animals were not only known to the general public, but they were regarded with a sense of awe, and viewed in an almost magisterial light by the readers of *Natural History.* And they were found only in the Old World, putting the reader in the proper mindset for thinking of the Old World as both different and inherently superior to the New. Buffon also used his rhetorical skills to that effect in another subtle way. He could have approached his topic by emphasizing those creatures that existed only in the Old World— what the reader might expect from a treatise entitled "Dissertation on Animals Peculiar to the Old World"—but, instead, Buffon's emphasis was on what *could not* be found in the New World, again calling attention to the fact that something must be amiss there. "In America," Buffon tells his readers, "Nature . . . [has] adopted upon a smaller scale."[51]

Buffon next provided the first clue as to why creatures in the New World degenerate. The magnificent creatures he listed as found only in the Old World—the elephant, lion, tiger, and so on—"require a hot climate for propagation." None of them could be found in America, and "this general fact is too important not to be supported by every proof."[52] Warm climates led to large, grand creatures, and Buffon judged all of America, especially North America, to be a cold continent. What's more, this passage contains an important hint about Buffon's ideas on geography and degeneration. The New World was inhabited by creatures that had migrated there, in Buffon's opinion, via a land bridge that had once connected Eurasia to North America. Buffon believed that compared to the land masses of the Old World, North America was a relatively new continent that had not had time to heat up or dry out; in such a poor climate migrants were bound to be smaller and feebler.

It was not only migrants who had come to the New World of their own accord that were doomed to degenerate. Even domesticated animals brought to America were destined to a similar fate. Sheep transported from the Old World took hold in the New, but "they are commonly more meagre, and their flesh less juicy and tender than those of Europe."[53] The ass, though present in the New World, was of lesser stock than its Old World brothers. Dogs of the New World were "absolutely dumb," "perfectly mute," and in cold regions "they have decreased in size. . . . Thus, they have degenerated."[54] Buffon's one exception to domestic degeneration in the New World was not particularly flattering: "The hog has thriven the best and most universally [there]."[55]

Buffon's next treatise on degeneration, "Dissertation on Animals Peculiar to the New World," was no less damning of American wildlife. Buffon thought that all else being equal, New World animals should have thrived, because they had few enemies, but much space. But all else was not equal: not only were the New World animals smaller, but the number of species there was paltry. Straying from the title of his treatise, which was to focus only on what was peculiar about the New World, Buffon noted that "if we reckon that 200 species of animals are in the known world, we shall find that more than 130 of them belong in the Old Continent, and less than 70 to the New. . . . There will not remain above 40 species peculiar to and natives of America."[56] Faced with these sorts of numbers, Buffon, the natural historian, felt compelled to conclude that "animated nature, therefore is less active, less varied, and even less vigorous, for by the enumeration of the American animals we shall perceive, that not only the number of species is smaller, but that in general, they are inferior in size to those of the old continent."[57]

In his next treatise, "Dissertation on Animals Common to Both Continents," Buffon reminded the reader of *Natural History* that some Old World species can endure cold and are found in North America. As such, it was critical to the degeneracy hypothesis to compare them in both the Old and New Worlds. And so on the very first page of this treatise, Buffon informed the reader that "the bears of Illinois, of Louisiana, etc., seem to be the same with ours; the former being only smaller and blacker,"[58] and quickly after that, "The wolf and fox are common to both continents . . . but all of them are smaller than those of Europe, which is the case with every animal, whether native or transported."[59] Domestic animals in the New World were just as vulnerable, "all the animals which have been transported from Europe having become less, and also those common to both continents being much smaller in America than those of Europe."[60] The only creatures in this New World that seemed to thrive there, as opposed to the Old World, were the cold-blooded ones: "Though Nature has reduced all the quadrupeds of the new world, yet she has preferred the size of reptiles, and enlarged that of insects."[61]

Besides the cold, there was another critical force fostering degeneracy in the New World, something that Buffon saw opposing "the aggrandizement of animated Nature" and "the formation of the principles of life"—a force that led life "on this vacant land" to become "shriveled and diminished."[62] And that force was humidity, particularly the humidity that was associated with swamps.[63] Buffon had a deep-seated fear of stagnant waters and swamps and the miasmas that were believed to arise from them. America, Buffon believed, had been under water for a far longer period than the Old World, and thus had not had sufficient time to completely dry out. Without proper drying out time, the land was full of stagnant swamps; it was extraordinarily humid, and that, combined with it being a world where "every thing concurs to diminish the action of heat,"[64] led to a condition in which "every thing languishes, corrupts, and proves abortive."[65] Everything except insects and reptiles, which flourished under conditions of high humidity and cold temperatures.

The last of Buffon's four treatises ("Treatise of Degeneration of Animals") served as a summary and does not include much new material on this topic. By this point in *Quadrupeds,* Buffon had exhausted himself and the reader with arguments about the degeneracy of New World animals. He rightly surmised that had he not convinced his readers by this juncture, there was little else he could do to sway them.

Buffon was certain that New World creatures were meager and weak compared to those found in the Old World, and he told the readers of *Natural History* that degeneracy in animals was a universal truth in America. We are left to wonder how Buffon, who had never been outside Europe, who displayed a healthy skepticism of his information sources, and who understood that variation defined nature, could make such a sweeping claim about animal degeneracy in the New World.

The answer lies in understanding Buffon's ideas on probability and what he called "moral certitude"; ideas that he formally presented in his article "Essay on Moral Arithmetic." In this essay, Buffon addressed the following question: at what frequency is an event so rare that one can reasonably and safely ignore it? To make this question more concrete, Buffon cast it in terms of the probability of dying at a given point in time, and whether this probability could safely be ignored.

Using mortality charts, Buffon calculated that the chance that a fifty-six-year-old man would not live out the day was about 1 in 10,189. He believed that despite these odds, "all men of that age, when reason has matured and experience has taught them all it can teach, have nevertheless no fear of death within 24 hours." Based on this, and other similar examples, Buffon came to the conclusion that "probabilities equal or less than this number should be discarded." But Buffon was willing to extrapolate well beyond the case of the potential mortality of a fifty-six-year-old man, "and so, in all games, wagers, risks and chances, and in every case, in a word where the probability is less than $1/10000$, it ought to be, and in effect is for us, absolutely negligible. . . . It gives us . . . more complete moral certitude."[66]

With respect to the case of degeneracy, what is critical is that Buffon believed that his ideas on probability and moral certitude could also be extended to *any natural phenomenon.* "When we observe a novel phenomenon, an effect in Nature still unknown," Buffon noted, " . . . as soon as it shall have occurred 13 or 14 times in the same fashion, we already have a degree of probability equal to moral certitude that it will occur a 15th time in a like fashion. At this point we readily make a broad assumption, and conclude from analogy that this effect depends on the general laws of nature . . . and that there is a physical certitude that it will always take place in this fashion."[67]

To see how Buffon came to this conclusion, we need to think in terms of a choice between two possible outcomes: either a New World animal is degenerate compared to the case of the same animal in the Old World, or it is not.[68] By chance alone, when we compare our first species in both environments, we would expect it to be degenerate in the New World with a probability of $1/2$. The odds that the first two animal species examined are both degenerate in the New World are $1/2$ times $1/2$, or $(1/2)^2$, the odds that the first three species are degenerate in the New World are $(1/2)^3$, and most importantly for our example, the odds that the first thirteen species are degenerate in the New World equals $(1/2)^{13}$ or 1 in 8,192, and in the case of fourteen species the odds are $(1/2)^{14}$, or 1 in 16,384 that all fourteen species will be found to be degenerate in the New World. In other words, somewhere between the 13th and 14th consecutive occurrence, we pass Buffon's moral certitude balance point of 1/10,000. Buffon's comparison of the Old and New Worlds was based on information provided by many more than fourteen different sources (often travelers), and these sources all indicated to him that animal life in the New World was degenerate. And so Buffon was morally certain that his degeneracy claim was correct. And, though his sources were far less numerous for indigenous human populations in the New World than for animals, Buffon was no less confident that Native Americans, too, were degenerate, and for precisely the same reasons that held true for animals.

As was true for his arguments regarding animal species, readers were prone to listen to what Buffon had to say about humans. Well before his degeneracy argument about New World humans, Buffon had established himself as one of the founding fathers of anthropology.[69] A large portion of the second volume of *Natural History* was dedicated to this subject, which included Buffon's argument that humans, rather than being a group of related but different species, each with a unique origin, were instead a single species.[70]

Buffon the anthropologist could be both poetic and harsh. In *Natural History,* Buffon took his readers on what his biographer, Jacques Roger, calls a "vast ethnographic tour of the world," covering societies in Europe, Asia, Africa, and the Americas.[71] Buffon most often began his description by outlining the physical attributes of the people in question—skin color, height, type of hair, shape of the lips, nose, face, and eyes, and so forth. In the course of such descriptions, he liberally mixed in his own aesthetic assessments: The Hottentots possessed "a stupid or ferocious look; hairy ears, body and limbs; skin as black as hard or tanned leather . . . long and limp breasts, the skin of the stomach hanging to the knees."[72] The Lapp women were ugly, and the Kalmucks were frightening. On the other hand, the Circassian women of Russia "have the most beautiful complexion and the most beautiful color in the world."[73] And Buffon's personal assessment of the human groups he described was not limited to their anatomical traits, but included their "habits" as well. The Yeco of Northern Japan, for example, displayed "savage" behaviors, ran around naked, and did not possess the slightest sense of right and wrong.[74]

Above and beyond specific case histories, Buffon searched for pattern. And, climate, it turned out, allowed him to decipher what he believed to be two general anthropological patterns. The first of these patterns arose from a comparison between people of the "hills" and people of the "plains." Inhabitants of the hills were "agile, full of energy, well built, spirited . . . the women there are generally pretty," while in the plains, "where the earth is crude, the air thick, and the water less pure, the countrymen are course, heavy, badly built, stupid and the countrywomen are almost all ugly."[75] Although this comparison was made long before his theory of American Indian degeneracy, Buffon had already begun his obsession with the effects of climate.

The second pattern Buffon discerned centered on climate and skin color. Because humans were all part of a single race, Buffon believed that skin color was, in part, a direct result of climate. Africans, for example, were dark-skinned, but they would become light-skinned if they were moved to northern climates for a few hundred years. Buffon went so far as proposing a direct test of this hypothesis: "To put the change of colour in the human species to the test of experiment, some Negroes should be transported from Senegal to Denmark, where the inhabitants have generally white skins, golden locks, and blue eyes, and where the difference of blood and opposition of colour are greatest." Then, in order to remove the effects of racial mixing on skin color, Buffon suggested that "these Negroes must be confined to their own females, and all crossing of the breed scrupulously prevented. This is the only method of discovering the time necessary to change a Negro into a White, or a White into a Black, by the mere operation of climate."[76] From there it would become the anthropologist of the future's job to see what sorts of changes would take place in these African Danes. As with the "hills" vs. "plains" pattern, what is most important about this example is that it shows the profound impact climate had in Buffon's eyes—effects that doomed the Native people of America.

Buffon's most damning comments about American Indians are found in his four treatises on New World degeneration.[77] There, Buffon claimed that American Indians "existed as a creature of no consideration in Nature," and railed that they were

a kind of weak automaton, incapable of improving or seconding her [Nature's] intentions. She treated them rather like a stepmother than a parent, by refusing them the invigorating sentiment of love, and the strong desire of multiplying their species. For, though the American savage be nearly of the same stature with men in polished societies; yet this is not a sufficient exception to the general contraction of animated Nature throughout the whole Continent. In the savage, the organs of generation are small and feeble. He has no hair, no beard, no ardour for the female. . . . He has no vivacity, no activity of mind. . . . He remains in stupid repose, on his limbs or couch, for whole days. . . . They have been refused the most precious spark of Nature's fire: They have no ardour for women, and, of course, no love to mankind. . . . Their love to parents and children is extremely weak. The bonds of the most intimate of all societies, that of the same family, are feeble; and one family has no attachment to another. . . . Their heart is frozen, their society cold, and their empire cruel. They regard their females as servants destined to labour, or as beasts of burden, whom they load unmercifully with the produce of their hunting, and oblige, without pity or gratitude, to perform labours which often exceed their strength. They have few children, and pay little attention to them. They are indifferent, because they are weak.[78]

In Buffon's eyes, nature had handled American Indians as it had other species in the New World—by the process of degeneration. But there was more to it than that, for Buffon attributed degeneration in New World animals in part to the action, or rather the lack of action, of American Indians. The failure of the Native Americans to prevent degeneracy in animals had two causes: first, there were simply too few indigenous people in America, and the numbers present were not sufficient to tame nature. This, though, was hardly the "fault" of the indigenous people.

What was their fault, in Buffon's assessment, was that the Indians were both too lazy and stupid to even try to control nature—they had failed to tame the land, and, in particular, they had failed to drain swamps and remove stagnant sources of water. This would increase humidity, and hence the rate of degeneration. The Indians had never even attempted to free themselves or other animals from the pernicious effects of degeneration. "In these melancholy regions," Buffon wrote, "Nature remains concealed under her old garments, and never exhibits herself in fresh attire; being neither cherished nor cultivated by man, she never opens her fruitful and beneficent womb."[79]

When control of the land was in better hands, Buffon believed the future of North America was more promising. He hoped that "in several centuries, when the earth has been tilled, the forests cut down, the rivers controlled, and the waters contained, this same land will become the most fruitful, healthy, and rich of all, as it is seen to be already in the parts that man has cultivated."[80] For the near future, though, degeneracy was the hallmark of all life in America. Buffon clung to his American degeneracy hypothesis—as it applied to both animals and indigenous Americans—until close to the end of his life. Only in 1778, at age seventy-one, in his fifth "Supplement" to Natural History, did he veer away from the New World degeneracy argument, and then only very quietly. Until this supplement was published, Buffon had applied the argument to all of America, both North and South. In the fifth supplement, however, Buffon restricted his discussion of degeneracy to South America, and in so doing, resuscitated life in North America by omission.

It is not clear why Buffon changed his views and restricted degeneracy in the New World to South America. It may, in part, have been due to an article published by Hughes Williamson in the first volume of *Transactions of the American Philosophical Society* in 1771.[81] In this article, which Buffon would surely have seen, as he himself was a member of the American Philosophical Society,[82] Williamson argued that transplanted Europeans—that is, the Revolutionary War generation—had already fundamentally modified the climate of North America for the better. Agricultural advances would make the climate more temperate and less humid, and as Williamson pointed out, Benjamin Franklin had long ago shown that transplanted Europeans were already experiencing very high rates of population growth.

None of this spoke well for North American degeneracy, and Buffon may have felt that he could drop the matter, and drop it safely, as Williamson's argument did not actually disprove the degeneracy argument. Instead, transplanted Europeans had simply created a new environment in which degeneracy was less likely to apply.

But, it was all too little, too late. The supplements were not read with the enthusiasm of the main volumes of *Natural History*, and even those who did read them would not necessarily have noticed that Buffon had changed the target of the degeneracy argument from all of the Americas to just the Southern continent.

In his fascinating book, *The American Enemy: The History of French Anti-Americanism*, Philippe Roger argues that French anti-Americanism "was born and proposed in philosophical circles" and that these philosophical circles revolved around Count Buffon.[83] It is not hard to see why Roger comes to this conclusion. Aside from a bit of silent backtracking toward the end of his career, Buffon—one of the best known names in Europe—was relentless in promulgating his theory of American degeneracy. No exceptions to degeneracy (except nasty insects and reptiles) were permitted, and Buffon's rhetorical skills made the case all the more convincing. Just as important, with Buffon as its leading advocate, supporters of the degeneracy hypothesis could legitimately claim a scientific framework for Old World superiority.

While a naturalistic anti-Americanism began with Buffon, it certainly did not end with him. Buffon, in fact, was reserved, compared to some who would follow. For while Buffon limited his arguments of degeneracy to animals and American Indians, his intellectual descendents—people like the Abbé Raynal, Abbé Cornelius de Pauw, and William Robertson—saw degeneration in the New World as all encompassing, could think of no reason to limit its scope, and readily extended the case to all Americans, including transplanted Europeans and their descendants.

CHAPTER THREE

"Noxious Vapors and Corrupt Juices"

The Europeans who pass into America degenerate, as do the animals: a proof that the climate is unfavorable to the improvement of either man or animal. The Creoles, descended from Europeans and born in America . . . have never produced a single book. This degradation of humanity must be imputed to the vitiated qualities of the air stagnated in their immense forests, and corrupted by noxious vapours from standing waters and uncultivated grounds.

CORNELIUS DE PAUW, 1768.[1]

The Prussian clergyman Cornelius de Pauw had much more to say along these lines in his 1768 book, *Philosophical Researches on the Americans (Recherches philosophiques sur les Américains)*. And de Pauw was not alone in extending Buffon's degeneration argument from animals and indigenous peoples to the more general case of Europeans and their descendants in the New World. The degeneracy argument that Buffon presented in *Natural History* would be picked up and expanded by other European writers, who were often all too happy to buttress claims of Old World superiority. For this vociferous group—whose leading spokesmen were the Prussian de Pauw and the French Abbé Raynal—Buffon's claims had not been bold enough. If the Count's arguments about climate and degeneracy applied to animals and Native Americans, then de Pauw and Raynal could see no logical reason to exclude Creoles—Europeans born in America—from the crippling effects of life in the New World.

Cornelius de Pauw was born in Amsterdam in 1739, a descendant of a well-to-do merchant named Michael de Pauw.[2] Michael was one of the directors of the Dutch West Indies Company, and he emigrated to North America in 1630. He quickly became owner of what is now Staten Island, but for reasons not known, he eventually returned to Europe, where he and his family remained.

Cornelius, the next de Pauw for which any significant data exists, was orphaned at a young age and sent to live with relatives in Liège. There the canon of a local cathedral took him in, and eventually arranged for de Pauw to receive an education at the University of Gottingen.[3] After his university training, he accepted a clergy position in the Prussian district of Xanten, and in March 1765 became subdeacon of a church there.[4] From then until his death in 1799, de Pauw led a cloistered life that appears to have been interrupted only by two visits to Frederick the Great at his palace at Potsdam.

In late 1767, while on a diplomatic mission to Berlin, Cornelius met the emperor of Prussia. Frederick was taken by the young man and offered him a position as his "private reader." De Pauw served in this role for a few months, but missing the seclusion he enjoyed in Xanten, he returned to his parish. Cornelius spent additional time with Frederick in 1775, but again returned to Xanten in short order. His stays with the emperor, brief though they were, immensely increased his standing in philosophical circles, and his works are often referred to as being penned by Frederick the Great's "personal advisor."

It was in his early years at the church in Xanten that de Pauw wrote *Philosophical Researches on the Americans*—a book that was quickly translated into German (1769), Dutch (1771–72), and eventually English (1789),[5] and one that went through eleven editions before the end of the eighteenth century.[6] Early on in *Philosophical Researches*, de Pauw rehashed the climate-based explanations of degeneracy put forth by Buffon—whom de Pauw often referred to as "our author"—and reminded his readers that "in the countries temperate in climate, and rich in vegetables, society has been established infinitely sooner than in the cold and barren."[7]

America, naturally, was one of those cold and barren places, a land "covered by immense swamps, which render the air extremely unwholesome."[8] Equally debilitating in de Pauw's eyes (as with Buffon) were the unhealthy consequences of the humidity: "The great humidity of the atmosphere, the prodigious quantity of stagnant waters, the noxious vapors, corrupt juices, and vitiated qualities of the plants . . . will account for that feebleness of complexion, that aversion from labour, and general unfitness for improvement of every kind, which prevented the Americans from emerging out of savage life."[9]

De Pauw believed that the effects of humidity and cold were even more damning and inescapable than did Buffon. While the Count saw a direct link between cold, humidity, and degeneracy, de Pauw saw both direct and indirect links. The cold and humidity of America tainted the soil, and as a result, a corrupted food chain was set in motion. Plants in the tainted soil provided poor nutrients for the animals, who in turn were degenerate food sources for humans. De Pauw saw degeneracy seeping from the air and waters and working its way straight up to humans: "The same ill qualities of the air . . ." he wrote, "are probably the true origins of the degeneracy in men and animals; as the same corrupt juices which infect the vegetable nature, must taint the blood, and subdue the powers of the animal."[10] From the "prodigious number of poisonous vegetables"[11] to the debilitating effects on humans, degeneracy permeated every nook and cranny of the New World.

De Pauw and Buffon did part company on one aspect of the cold and wet world of North America. Buffon thought America was a newer landmass than the Old World and simply had not had time to dry and warm. De Pauw, on the other hand, thought the New and Old Worlds to be the same age, but surmised that America had been hit by a huge flood that was unique to the New World. America was just as old as other areas of the world, then, but parts of it had been under water more recently. As evidence for this, de Pauw claimed that Native American tales told of "ancestors [who] were forced to betake themselves to the mountains at the time of a mighty flood."[12] But this distinction about the origins of America's dampness hardly mattered in terms of documenting the *effects* of degeneracy, which were the focus of de Pauw's attack on America.

While Buffon spent the majority of his text on animal degeneracy, de Pauw's primary interest was in humans. Indeed, de Pauw's remarks on animal degeneration are few and far between. He did briefly touch on the diminished size of American animals when compared to those of the Old World, and de Pauw, like Buffon, saw the ill effects of climate leading to "the prodigious propagation of insects, venomous serpents and infected vegetations, which so unhappily distinguish this hemisphere."[13] For good measure, de Pauw added accounts he had read of giant frogs in Louisiana that uttered sounds not unlike that of the cow.

It was when he turned to humans—both indigenous and Creole—that de Pauw ramped up his rhetoric, and was even more vitriolic than Buffon. When he wrote about the American Indian, de Pauw's language was filled with comments on physical and moral corruption. Indian males were hairless and resembled eunuchs. These effeminate male savages had "milk, or a kind of milky liquid in their breast," and every one of them was "tainted with venereal disease."[14] The women were no better off—they too were infected with venereal disease, and though their process of childbirth was painless, they were not as fecund as European women—a clear sign, in de Pauw's eyes, of "a derangement of constitution."[15]

While de Pauw thought it impossible to know whether or not Indians were happy, he felt that their degenerate environment had made them undeniably lazy—"their hatred of labor," he wrote, "never induced them to cultivate the earth."[16] In addition to laziness, the indigenous Americans were presented as completely lacking a sense of compassion. As a case in point, de Pauw noted that although Indians respected their elders, as soon as those elders became ill, "they become an encumbrance." Indians then reverted back to their animal nature, de Pauw told his readers, and abandoned the same elders they recently respected, like "beasts of prey, who are left to perish when they are no longer able to hunt and provide for themselves."[17]

De Pauw was fond of describing the Indian as "vegetating" rather than living, and noted that "a stupid imbecility is the fundamental disposition of all Americans."[18] The Indians were incapable of forethought, lacked any "nobility of mind," and were "useless to themselves and to society."[19] All of this, of course, was a consequence of the direct and indirect results of having had the misfortune of living in the horrid, cold, and damp environs of the New World.

Above and beyond de Pauw's vitriol, however, some of his descriptions of American Indian life bordered on the absurd. Eskimos, for example, "like some quadrupeds, lick their new born infants."[20] As an even more extreme case, de Pauw described an (unidentified) Indian culture where "mothers take the heads of their infants, three or four days old, in hand, and begin to fashion them into the form of a pyramid, a cone, cylinders some to be quite flat, others an exact square; and the last, which is the completion of beauty, perfectly round."[21] As was the case for Buffon's anthropology and natural history, these sorts of descriptions raise a question to the modern reader: since de Pauw never left Europe, let alone traveled to America, from where did he get these incredible tales about American Indians?

De Pauw himself was sensitive to this question. "We must confine ourselves to facts," he tells the reader of *Philosophical Researches*, "lay them open such as they are, or one believes them to be, without hatred, without prejudice, without respect, except for the truth."[22] De Pauw presented himself as a man in possession of a healthy dose of skepticism regarding the veracity of his sources, noting that "one may lay it down as a maxim, that out of one hundred [travelers], there are fifty that are liars, not through interest, but through ignorance, thirty through interest or pleasure of imposing on the public, and about ten who are honest and aim at truth."[23] But the historical record shows that this was a false veneer, and that de Pauw consistently accepted accounts that suggested American degeneracy, and ignored any data to the contrary.[24]

It was among these sorts of wild stories about American Indians that de Pauw penned his most controversial remarks—those regarding the degeneracy of Europeans in the New World. It was one thing to expand on Buffon's ideas on American animals and indigenous people; it was quite another to suggest that Europeans who were born in, or traveled to, the New World would degenerate along the very same lines as all other life forms in that godforsaken wasteland.

Europeans born in America—Creoles—were a special target of de Pauw. Stuck in a world "corrupted by noxious vapours from standing waters and uncultivated grounds,"[25] these hybrid New–Old Worlders would suffer the same fate as the indigenous people that surrounded them. Though he never explained why, de Pauw saw degeneracy acting in a slightly different way when it came to Creoles. Indians were degenerate from birth, but not so the Creoles, who actually started out life with an advantage compared to Europeans of the Old World: "The Creoles both of South and North America come to a maturity of intellect . . . more early than the children in Europe,"[26] de Pauw informed his readers. But the "true" European reader of *Philosophical Researches* need not fear, for de Pauw followed this surprising comment with his typical damning prose: "This anticipation of ripeness is short-lived, in proportion to the unreasonableness of its appearance; for the Creole falls off, as he approaches puberty."[27]

This descent toward degeneration was rapid: the Creole's "vivacity deserts him, his powers grow dull, and he ceases to think at the very time that he might think to some purpose."[28] It took a bit longer with the Creoles, but in the end, de Pauw argued, they could no more escape the effects of degeneracy than animals or Indians. Unless, of course, they changed the climate itself, by, for example, decreasing the amount of stagnant water. Indeed, de Pauw recognized that the Creoles had made some progress on this front, "yet the change in climate has not been so great as might have been expected, owing to vast regions covered with water and woods surrounding the spots which have been cultivated; nor has the degeneracy in men and animals of European origin diminished in the proportion that was expected."[29] The prospects for mitigating the effects of climate, and hence degeneracy, were, in de Pauw's eyes, slim, and the odds were that "the people of whom we are speaking will never rise above their present abject condition."[30]

As with the Indians, the physical effects of degeneration in Creoles were accompanied by mental ones. Creoles had no artistic abilities worth mentioning—"even at this day," de Pauw noted, "all the Americans and Creoles united cannot produce a picture fit to be in the collection of an Alderman."[31] And their scholastic skills were no greater: it was obvious to de Pauw "that the professors of the University of Cambridge, in New England, have not formed any young Americans to the point where they are able to bring them out into the literary world."[32] As for the promise of remedying this lack of scholars, the prospects were as dismal as they were for the physical effects of degeneration. So damning were de Pauw's words on Creoles, that historian Gilbert Chinard has argued that they even offended Buffon—no slouch himself when it came to degeneracy and vitriolic rhetoric. Chinard suggests that one of the reasons that Buffon may have silently backtracked on North American degeneracy in late life was an instinctive recoil from de Pauw's extension of degeneracy to those of European extraction.[33]

Given that Buffon had already eloquently and powerfully laid out the basic idea of degeneracy in North America, why did de Pauw feel the need to write a book-long extension of this idea, with an emphasis on Creoles? Part of the answer probably lay in the general fear that many Europeans had regarding the potential of the New World. While financial opportunities in this New World certainly awaited risk takers, the idea that Europeans could now emigrate and obtain vast tracts of land on the other side of the globe was not necessarily appealing to Old World monarchs. Indeed, in his book, *Mirage in the West*, Durán Echeverría has suggested that de Pauw wrote *Philosophical Researches* in an attempt to ingratiate himself with one such monarch—Frederick the Great.[34]

Frederick had a strong antiemigration policy when it came to America, going as far as establishing a special agency in Hamburg whose sole function was to prevent emigration to the New World, and instead to attract potential newcomers to Prussia. After having so warmly welcomed de Pauw into his court, Frederick may have received an elaborate "thank you" from de Pauw in the form of *Philosophical Researches*. In any event, it is clear that de Pauw's relationship with Frederick dramatically increased the popularity of his book, in part accounting for its numerous editions and translations.

This special relationship with the emperor may also have been one of the reasons that, to the ultimate dismay of the Revolutionary War generation of Americans, de Pauw was given the honor of composing the entry on America in the influential *Supplément de l' encyclopédie* in, of all years, 1776.

Cornelius de Pauw was a one-dimensional character in the story of degeneracy. His vitriol was genuine, but it was his only historically defining trait. His sole claim to fame (or infamy) is his work on New World degeneracy per se, which is to say that he was read for his thoughts on degeneracy and not for historical content. This was not the case for the French Abbé Guillaume-Thomas Raynal, who could be equally vituperative when it came to America and degeneracy, but who was a deeper, more respected, but also more paradoxical, character than de Pauw.

Raynal published an extremely popular set of books entitled *A Philosophical and Political History of the Settlements and Trade of the Europeans in the East and West Indies*.[35] These volumes were packed with world history, economics, political philosophy and anthropology. When it came to the topic of New World degeneracy, Raynal picked up where Buffon and de Pauw left off. Indeed, perhaps the most famous, and arguably the most inflammatory, of all the passages on degeneracy in transplanted Europeans in the New World can be found in Raynal's work.

Guillaume-Thomas Raynal was born on April 12, 1713, in the Saint-Rouergue region of Southern France. His father was a local merchant of note, who increased his standing by marrying Catherine Girels, the daughter of well-to-do local nobility. Raynal was well educated in Jesuit schools, and attended the Jesuit College de Rodez, where he was eventually ordained a priest. Shortly after his ordination, he accepted a position as professor of philosophy at the Université de Toulouse. Sometime in 1747 Raynal left Toulouse to move to Paris, where he acted as a chapel priest, leading mass at the Church of Saint-Sulpice. Abbé Raynal's time at Saint-Sulpice did not go well, and he was apparently removed as chapel priest for selling "spiritual favors," and for the more egregious sin of daring to sell cemetery plots to Protestants.[36]

The events at Saint-Sulpice soured Raynal on the life of the clergy, and from that point on, his life became one of author and philosopher-at-large. With the help of the French foreign minister, Raynal secured a position as editor of the powerful *Mercure de France* newspaper, where he established important literary connections.[37] He also sharpened both his rhetorical and debating skills to such an extent that he would eventually boast that when in a heated debate with an opponent, it was enough for a

GUILLAUME THOMAS RAYNAL.

du Défenseur de l'Humanité, de la Vérité, de la Liberté.   ELIZA DRAPER.

3.1. Guillaume-Thomas Raynal, one of the most prominent advocates of the degeneracy theory. Engraving by Nicolas Delaunay, drawing by Charles-Nicolas Cochin.

rival to cough for him to capture the momentum and verbally eviscerate his challenger.[38]

Raynal published a series of pamphlets and books, including *History of the Stadtholderate* (describing the Dutch struggles against colonial Spain), *History of the Parliament of England,* and *Spirit of the Laws of England.*[39] These works, though not nearly as popular or sweeping in nature as Raynal's *Philosophical and Political History*, established him as an author and philosopher, and opened doors for him at the very best salons in Paris. All of these works allowed Raynal to hobnob with, and befriend, many of the intellectual giants of his day, including Rousseau and Benjamin Franklin.[40]

The first edition of the six-volume *Philosophical and Political History* was published in Amsterdam in 1770, and then in Paris in 1772.[41] The second edition appeared in 1774, and a third edition, with a much more detailed analysis of the American Revolution, appeared in 1780. Over the course of the following three decades, these volumes were translated into English, German, Dutch, Italian, and Spanish and went through at least thirty editions and revisions.[42] This work was widely read across Europe, and Napoleon would one day claim to be a disciple of Raynal, carrying *Philosophical and Political History* with him during his campaign in Europe.[43] Raynal's work could also be found in the libraries of many of the most influential colonial American plantation owners.[44]

Philosophical and Political History has the feel of an encyclopedia, with multiple entries on a wide array of historical and economic topics. And indeed, although Raynal wrote the majority of the text in these volumes, and is rightly credited with "authorship," he also played the role of editor and organizer for the many sections that were written by his collaborators.[45] The most famous and most prolific of these collaborators was Denis Diderot, an author and playwright, who was simultaneously editing his own massive twenty-eight-volume *Encyclopedia of Arts and Sciences.*[46]

Raynal and his team drew their information from many sources, including the accounts of travelers, government documents and statistics, colonial administrators, foreign consultants from Denmark, England, Holland, Portugal, and Spain, Buffon's *Natural History*, the work of de Pauw, and publications from the American Philosophical Society.[47] Although Raynal was not above the occasional act of plagiarism from these sources, most often he would choose a topic and either write on it himself, or assign it to a collaborator, and then edit what he received to create a continuity to the thousands of pages that made up *Philosophical and Political History.*

Raynal's magnum opus was stuffed with information, including an economic, political, and societal history of all the countries that had been settled by Europeans—most especially those settled/conquered by the Spanish, English, Portuguese, Dutch, Prussians, Danes, Swedes, and Russians—and often including historical information that predated European settlement by centuries. The text of *Philosophical and Political History* has a passionate feel to it, with Raynal assuring his readers: "When I began this work, I took an oath that I would adhere strictly to truth; and hitherto I am able consciously to declare that I have not departed from it. May my hand wither if that should happen."[48]

One reason for his passionate tone is that Raynal was possessed by the New World, and the reader is introduced to this obsession on page 1 of volume 1, where the author proclaimed that "no event has been so interesting to mankind in general, and to the inhabitants of Europe in particular, as the discovery of the New World." Raynal chose his words here carefully—the discovery was *interesting*. The question—the one that Raynal desperately wanted to answer—was whether, all things considered, this interesting discovery had been advantageous or disadvantageous for humanity. His position on this question evolved through the numerous editions of *Philosophical and Political History,* beginning with a negative slant, and ending with a slightly positive one.

In the later editions of *Philosophical and Political History*, Raynal concluded that as a whole, the discovery of the New World was a good thing, but it was a razor-thin decision, and one heavily influenced by the promise of liberty and freedom raised by the American Revolution. Without that revolution, he probably would have concluded that this discovery resulted in a net loss for mankind—in fact, that is exactly what he did conclude in the 1772 and 1774 editions of *Philosophical and Political History* that predated the American Revolution. In these early editions, transplanted Europeans residing in North America were caricatured as ignorant and degenerate.

Raynal's flip-flopping about the value of the New World was reflected in a contest that he sponsored at the Académie de Lyon sometime in the 1780s. Putting up a prize of 1,200 livres, Raynal solicited essays on the question "Was the discovery of America a blessing or a curse to mankind?" Although forty-four essays were submitted, and the contest lasted for years, the judges failed to agree on a winner.[49] Paralleling Raynal's own decision-making process, of the eight essays that survive, four suggest "advantageous" and four not.[50]

Although Raynal was ultimately convinced of the beneficial effects associated with the discovery of the New World, much of the text in *Philosophical and Political History* is devoted to the downside of that discovery. For Raynal, the negative consequences associated with the founding of the New World were linked to a pernicious combination of colonialism and American degeneracy. Raynal saw colonialism as the work of the devil himself, and the New World was all about European colonialism. The opening of this world led to colonial powers violently oppressing those in the New World, raping the land and the women, and securing whatever resources—especially gold and silver—they could get their greedy hands on. Such conquests would never have happened had the New World not been discovered.

Of the horrors inflicted by colonial powers, Raynal saw none worse than slavery. Indeed, Raynal was arguably the most vocal of all enlightenment philosophers in condemning slavery. Dozens of pages in *Philosophical and Political History* are devoted to denouncing slavery, whether the slave was a Negro or an Indian. Raynal, for example, was so pained by the Spanish enslavement of South American Indians that he lamented, "O God! Why did'st thou create man.... Thy prescience certainly foresaw the atrocious acts which the Spanish were to commit in the New World!"[51]

In Raynal's eyes there was plenty of blame to go around for why colonial Europe had so brutally assaulted the New Worlders. Of course, greedy merchants were involved, but so too were the clergy, who in Raynal's eyes, were more than happy to have new flocks to preach to, no matter how obtained, and were indirectly responsible for the evils of colonialism. The rest of the blame could be laid at the footsteps of tyrannical leaders—hereditary monarchs for the most part—who financed colonial adventures and reaped the benefits.

Not surprisingly, Raynal's rabid anticlerical and antimonarchical writings did not go over well with some in France. The church labeled Raynal's work as "impious, blasphemic, seditious, tending to incite the people against the sovereign authority,"[52] and he was censured by the Faculty of Theology at the Sorbonne. His books were placed on the Catholic Church's Index of Forbidden Books,[53] and the Crown banned the first and second editions. For good measure, they not only banned the third edition as well, but in 1781 had it burned by the public executioner[54] and put a warrant out for Raynal's arrest, though he was given sufficient notice to leave France before any such arrest was attempted.[55]

European greed was not the sole reason that Raynal believed that the discovery of America was a curse. The New World's troubles were not all external. True, conquest-minded enemies had ravished the New World, but there was a deeper, more fundamental problem in America that couldn't be blamed on conquerors, and that was the inherent disease of degeneracy.

On the question of climate in the New World and its role in degeneracy, Raynal turned to Buffon's *Natural History* for answers, for as he saw it, "Barbarous soldiers and rapacious merchants were not proper persons to give us just and clear notions of this hemisphere. It was the province of philosophy alone . . . [which allowed] us to see America such as nature has made it."[56] And in the realm of philosophy as it relates to the nature of life, Buffon was *the* source.

Parroting, but almost never citing, Buffon, Raynal reinforced the view that life in the New World was puny compared to that found in the Old: "We see more species of animals by two-thirds, in the old continent than the new; animals of the same kind considerably larger. . . . Nature seems to have strangely neglected the new world."[57] The reason for this degeneracy of course was the horrid climate of the New World, which had "remained with the waters of the sea a much longer time than the old."[58] This more recent submersion, this "infancy of nature,"[59] led to an accumulation of the stagnant waters and "malignant vapours" that had forced Buffon, de Pauw, and now Raynal, to doom the New World to degeneracy. But Raynal went one step further. The degeneracy that resulted from "damper air and a more marshy ground" was akin to a disease that "infected the first principles of the subsistence and increase of mankind."[60] It was in this mindset that Raynal made one of the more outlandish suggestions so far about degeneracy—that the Salem witch hunts in seventeenth-century New England were a result of "vapours and exhalations of a soil newly broken up."[61] Degeneracy not only led to smaller life forms, but to delusional behavior, and perhaps even powers of black magic.

Given his strong position on the damning effects of degeneracy through climate, Raynal's writings on New World American Indians are somewhat surprising. In fact, in *Philosophical and Political History*, Raynal was pulled in opposite directions on Indians and degeneracy, appearing at points to contradict himself. He had made much of the powers of degeneracy, and was forced to admit that they should apply to indigenous New World people as well as New World animals. But, Raynal also found Rousseau's ideal of the noble savage—man at peace with the land, uncorrupted by civilization in his natural state—appealing.[62]

Raynal's writing became convoluted when he attempted to accommodate both these positions. To allow for degeneracy but salvage the idea of the noble savage, Raynal adopted a tortured strategy: he would praise particular aspects of specific Indian societies while at the same time suggesting that, at the continental level, the damning effects of degeneration generally produced indigenous peoples that were far inferior to Old World Europeans.

When describing the Indian tribes of Canada, Raynal noted that although abstract terms were missing from their vocabulary, "no Greek or Roman orator ever, perhaps, spoke with more strength and sublimity than one of their chiefs."[63] What's more, because individuals in these Indian tribes did not accumulate possessions, Europeans had assumed some sort of fundamental social flaw existed in these societies—only barbarians would fail to try to better their lot in life. But Raynal read the situation very differently: "As the savages possess no riches," he told the reader of *Philosophical and Political History,* "they are of a benevolent turn. A striking instance of this appears in the care they take of orphans, widows and infirm people."[64]

One finds similar praise from Raynal when it came to specific acts in specific Indian tribes. But it was all for naught, for, as a whole, indigenous New World inhabitants were doomed to the effects of degeneracy. In the same volume in which readers learned of the benevolence of some Indians in some tribes, Raynal interspersed general comments on the ultimate power of degeneracy. "Colder blood . . . ," wrote Raynal of the American Indian, "[and] a constitution more phlegmatic by the dampness of the air and ground . . . doubtless blunt the irritability of the nervous system"—these savages were lacking in feeling and emotion.[65] This was no reason to enslave them, of course, but it did mean that they were degenerate. Raynal then reviewed de Pauw's evidence for degeneracy (referring to de Pauw as a "celebrated writer"):

> Love among the Americans is never productive of industry, genius and character, as it is among the Europeans. . . . The air and the climate, the moisture of which contributes so powerfully to vegetation, does not bestow upon them any great warmth of constitution. . . . The blood of these people is watery and cold, the males have sometimes milk in their breasts. . . . They have few children, because they are not sufficiently fond of women.[66]

The problem was deep within the New World, and not amenable to an easy fix, as it lay within "the same sap which covers the countries with forests, and the trees with leaves."[67] Raynal was inclined to de Pauw's view, but was willing to cede that Indians, "are much fonder of their children"[68] than Europeans, and that if men weren't fond of women, it wasn't their fault, as they were simply too exhausted from toiling to keep their family alive. The reader got the sense that Raynal wanted to praise the Indian—the noble savage—but he couldn't escape Buffon and de Pauw's degeneracy claims. In the end, for Raynal, "The ruin of that world is still imprinted on its inhabitants. They are species of men degraded and degenerated in their natural constitution. . . . A damper air and more marshy ground must generally have infected the first principles of the subsistence and increase of mankind."[69] That imprint was on the face of the Indian, but the cause of the imprint affected the Creole as well.

Raynal's position on degeneracy and transplanted Europeans and their descendants evolved over the course of editions of *Philosophical and Political History*. Across all editions, he had a soft spot for Quakers, particularly because of their opposition to slavery.[70] But aside from the Quakers, early editions of *Philosophical and Political History*—those written before the American Revolution—did not paint Creoles in a positive light. They, like the Indians, were subjected to the vagaries of degeneracy. "The Creoles," wrote Raynal, "though habituated to the climate since birth, are not as sturdy in their work, or as smart in conflict as the Europeans; that is to say that education has not prepared them, or that nature has made them weak. Under this alien (foreign) sky, their spirit is unnerved like the body."[71]

This unnerved Creole spirit, Raynal believed, had damning implications for the intellectual contributions one might expect from this group of New Worlders. And it was in that mindset that he uttered the single most famous statement about Creole degeneracy—a statement that would later infuriate Thomas Jefferson—namely that "one should not be surprised that America has yet to produce a good poet, a clever mathematician, a genius in even one art or science."[72] Creoles, Raynal decried, must be "happy . . . with mediocrity."[73]

Then in the 1780 edition of *Philosophical and Political History*, in a remarkable turnabout, Raynal not only apologized for his disparaging remarks about transplanted Europeans and their descendants, but attempted to explain why he was drawn to his earlier, incorrect opinion. "The inhabitants were universally thought to be less robust in labour, and less adapted to the arts than their ancestors," Raynal began, and " . . . it was concluded that they were degenerated, and unable to elevate their minds to any complicated speculations." But that, Raynal now claims, was an error. He opened his mea culpa:

> In order to dispel this fatal prejudice it became necessary that a Franklin should teach the philosophers of our continent the art of governing the thunder. It was necessary that pupils of this illustrious man should throw a striking light on several branches of the natural sciences. It was necessary that eloquence should renew in that part of the New World, those strong and rapid impressions that it made in the proudest republics of antiquity. It was necessary that the rights of mankind,

and the rights of nations, should be firmly established there, in original writings, which will be the delight and the consolation of the most distant ages.[74]

Raynal, then, was claiming that both the brilliance of Franklin per se and the wonders of the American Revolution convinced him to change his mind about Creoles as a product of degeneracy. The latter reason seems to be sincere; the former perhaps not.

At the time that the first and second editions of *Philosophical and Political History* appeared, the American Revolution had not yet begun. When Raynal mentioned conflict between the American colonies and England in these editions, he came down on the side of the colonies, but did not advocate complete independence, as he feared America might evolve into a tyrannical power herself. Between the second and third editions—that is, between 1774 and 1780—Raynal became a fervent supporter of the revolution and its ideals of liberty and freedom.[75] At that point he felt obliged to step back and apologize for his earlier disparaging remarks regarding Creole degeneracy in the New World.[76]

Raynal's other purported reason for recanting his previous remarks—the discoveries and advances made by Benjamin Franklin—is more suspect. Franklin's 1752 Royal Society paper on lightning had been on the books for nearly two decades when Raynal published the first edition of *Philosophical and Political History*. While it is always possible that the brilliance of Franklin's discovery did not become clear to Raynal until sometime after 1774, it is unlikely, and there is no evidence to support that claim. Indeed, Franklin and Raynal had also met on numerous occasions during the good doctor's visits to France in 1767 and 1769. One of these encounters is particularly telling. As Thomas Jefferson recalled the event many years later:

The Doctor [Franklin] told me at Paris the two following anecdotes of the Abbé Raynal. He had a party to dine with him one day at Passy, of whom one half were Americans, the other half French, and among the last was the Abbé. During the dinner he got on his favorite theory of the degeneracy of animals, and even of man, in America, and urged it with his usual eloquence. The Doctor at length noticing the accidental stature and position of his guests, at table, " Come," says he, "M. Abbé, let us try this question by the fact before us. We are here one half Americans, and one half French, and it happens that the Americans have placed themselves on one side of the table, and our French friends are on the other. Let both parties rise, and we will see on which side nature has degenerated." It happened that his American guests were Carmichael, Harmer, Humphreys, and others of the finest stature and form; while those of the other side were remarkably diminutive, and the Abbé himself particularly, was a mere shrimp. He [Raynal] parried the appeal, however, by a complimentary admission of exceptions, among which the Doctor himself was a conspicuous one.[77]

Raynal had the opportunity to talk with Franklin on the very issue of American degeneracy before the first edition of *Philosophical and Political History* appeared. Even after learning that Franklin thought the whole idea was absurd and had used his frontier wit to demonstrate it, Raynal published his condemnation of Creoles in the 1770 edition. His mea culpa was a function of the promise he saw in the revolution, and not the grandeur of Franklin. Once his admission of guilt was made, however, Raynal spared no praise for the burgeoning republic, noting that it was a land of "healthy and robust men, of a stature above the common size,"[78] a country that "cannot fail to become one of the most flourishing countries on the globe,"[79] and he fervently wished that "your duration, if possible, be long as that of the world!"[80]

And so, Raynal, like Buffon, pulled back from his initial attack on North America (de Pauw never did). In Buffon's case, the retreat was subtle, almost imperceptible to the reader, while with Raynal, the retreat—if the reader just happened to own a copy of the third edition or beyond—was more direct. But again, it was too little, too late. In both cases the change of heart was akin to a buried retraction (in supplements and later editions) of a front-page headline. The founding fathers—most especially Thomas Jefferson—would have none of that.

CHAPTER FOUR

"Not a Sprig of Grass That Shoots Uninteresting"

Claims about degeneracy and North America were well known to the founding fathers, and these accusations could not go unanswered.[1] So offensive was the notion of degeneracy that Thomas Jefferson, the most vociferous Francophile of all the founders, took it upon himself to refute Buffon and his supporters. From the early 1780s onward, this became one of Jefferson's great obsessions. The only book he ever wrote—*Notes on the State of Virginia*—is, in part, a treatise discrediting the idea of degeneracy.[2]

Jefferson was not the only founding father who worried about Buffon's claims of degeneracy. Benjamin Franklin, James Madison, Alexander Hamilton, and John Adams all weighed in on this issue, albeit to different extents, and in a variety of manners.[3] John Adams wrote how pleased he was that his colleagues were responding to "the mistakes of Raynal, and . . . Buffon, so unphilosophically borrowed from . . . those despicable dreams of de Pauw."[4] Despite the fact that Adams had his chronology backward—Raynal and de Pauw borrowed from Buffon, not the other way round—he too wanted the claims countered, so as not to impede American growth. Adam's wife, Abigail, also railed against assertions of degeneracy, mocking the claim that Americans were not "half so virtuous" as those in the Old World.[5]

James Madison was offended by degeneracy, both because he himself was something of a naturalist, and because he knew how much the inferiority claim rankled his mentor, Jefferson. In a long letter to Jefferson in June 1786, Madison provided his colleague with evidence to use in the

4.1. John Adams spoke of the "the mistakes of Raynal, and . . . Buffon . . . [and] those despicable dreams of de Pauw." Portrait by Gilbert Stuart.

case against Buffon.[6] After addressing various pressing political issues, and noting "I have a little itch to gain a smattering in chymistry," Madison opted to fill the remainder of his letter with the results of an examination he had made of a "minor quadruped," the weasel.

Madison went into detail in his account of the weasel, noting such attributes as "the lower jaw which was white for about 1/2 an inch back from the under lip," "the membrane of the bladder very thin," "the spleen was of the same color on both sides," and "its smell was a sort of rankish musk, but not so strong as to be very offensive." In addition to this in-depth prose description, Madison also enclosed a table with measurements on everything from the "width of the ears horizontally" to a measure of the "distance between the anus and the vulva."

More than mere description, Madison's enumeration of all things weasel was set in the context of a comparison to other closely related European species of mustelids (the family that contains the weasel) which Buffon had described in his works.[7] Indeed, Madison went as far as to draft a table which compared the American weasel with European species described by Buffon. And, at least for the single weasel from which Madison drew his measurements, the American species was as large as the European equivalents, leading Madison to assert that the data "certainly contradicts his [Buffon's] assertion that of the animals common to the two continents, those of the new are in every instance smaller than those of the old." Madison ended his letter somewhat anticlimactically by noting that Buffon "seems to have given up this point [American degeneracy] himself," in one of his supplements to *Natural History*. Nonetheless, de Pauw hadn't stepped down and neither had Raynal (at least with respect to degeneracy in animals), and most people, in fact, had no idea that Buffon had "given up this point."

What makes Madison's 1786 letter to Jefferson so interesting is not the excruciating detail on weasel biology, or even the table he provided—such things were common in natural history descriptions of the day. What stands out is that in the midst of writing letters that were focused on issues such as the proposition to hold a constitutional convention, whether paper money should be adopted, and what to do about an empty treasury,[8] Madison thought it important enough to devote some space to providing his friend with ammunition to counter the degeneracy claims coming out of Europe.

Benjamin Franklin's role in the degeneracy debate entailed more than a folksy, dinner-party response to Raynal's claims. His brainchild society—which would eventually be the American Philosophical Society—thought it to be so important for its members to know of degeneracy that it purchased an expensive copy of *Natural History* for all to read.[9] Franklin himself understood all too well that European emigration was vital for the future of America, and for decades had been collecting data on

4.2. Letter on degeneracy and the natural history of the weasel from Madison to Jefferson (June 19, 1786). Note the table on the right-hand side. Library of Congress.

both population growth and climate to demonstrate the true character of North America.

In his 1755 essay *Observations Concerning the Increase of Mankind, Peopling of Countries, etc.*, Franklin demonstrated how quickly the population of the colonies was growing—doubling every twenty-five years. The reason for the rapid growth, Franklin claimed, was "land being thus plenty, and so cheap as that a labouring man . . . can in a short time save enough to purchase a piece of new land sufficient for a plantation, whereon he may subsist a family." The consequence of such cheap land was that "marriages in America are . . . more generally early than in Europe," leading to faster growth in the former. But there was more to it than that.

It was not only the vast resources in North America, Franklin argued, that lead to growth: "The great increase of offspring . . . is not always owing to greater fecundity of nature, but sometimes to examples of industry in the heads, and industrious education, by which children are enabled to provide better for themselves, and their marrying earlier is encouraged from the prospect of good subsistence."[10] Many years later, these visions of North America—its bountiful nature, and its industrious citizenry— were often invoked to counter arguments of degeneracy.

Franklin's skills were also brought to bear on the question of degeneracy in another manner. The good doctor had been collecting data on the moisture of the air in America, England, and France since the 1750s, and in 1780 he was ready to pronounce that the moisture content in Europe was *higher* than in America, in direct opposition to Buffon's claim that America was degenerate as a result of its excessive moisture.[11] Franklin didn't gather this data in response to Buffon's claims, but it would prove useful in countering them.[12]

Alexander Hamilton understood that commercial business would be the key to America's future, and the theory of degeneracy was most certainly not good for business. The almost superhuman endurance that led Hamilton to write the lion's share of *The Federalist Papers* is well documented,[13] but what is often overlooked is that in the flurry that lead to the publication of this series of highly influential essays, Hamilton found the time to take a swipe at Buffon and de Pauw's ideas. Readers of *Federalist* number 11, "The Utility of the Union in Respect to Commercial Relations," might not have expected any reference to the theory of degeneracy, but the essay closed with Hamilton's thoughts on just this matter.

For Hamilton, trade was critical for America's success, and a strong union could only help on this front. Indeed, the future secretary of the treasury's goal was a United States with a strong enough trade base and a powerful enough federal government to "oblige foreign countries to bid

4.3. Alexander Hamilton. In *Federalist* 11, Hamilton mocked the "profound philosophers . . . [who] have gravely asserted that all animals, and with them the human species, degenerate in America." Portrait by John Trumbull.

against each other, for the privileges of our markets." Part of the battle to produce a powerful trade base was psychological—people had to be persuaded of America's great potential. To do this, Hamilton—who wrote *The Federalist Papers* under the pseudonym "Publius"—first described the sociopolitical landscape of the eighteenth century, at least as he saw it: "The world may politically, as well as geographically, be divided into four parts, each having a distinct set of interests. Unhappily for the other three," Hamilton continued, "Europe, by her arms and by her negotiations, by force and by fraud, has, in different degrees, extended her dominion over them all. Africa, Asia, and America, have successively felt her domination."

European power, Hamilton told his readers, led to corruption and to false feelings of grandeur: "The superiority she [Europe] has long maintained has tempted her to plume herself as the Mistress of the World, and to consider the rest of mankind as created for her benefit." The ideas of Buffon, Raynal, and de Pauw are then tackled directly in *Federalist* 11: "Men admired as profound philosophers have, in direct terms, attributed to her inhabitants a physical superiority, and have gravely asserted that all animals, and with them the human species, degenerate in America—that even dogs cease to bark after having breathed awhile in our atmosphere." Indeed, the only footnote in all of *Federalist* 11 is to this barking dog reference, which Hamilton correctly credits to de Pauw's *Philosophical Researches on the Americans*.[14]

The "arrogant pretensions" of men like Buffon and de Pauw, Hamilton declared , must be met head on. "It belongs to us," he told the *Federalist* reader, "to vindicate the honor of the human race, and to teach that assuming brother, moderation." But how? The answer to that question is the answer to almost all rhetorical queries posed by Hamilton's (and Madison's and Jay's) *Federalist* essays: "Union will enable us to do it. Disunion will add another victim to his triumphs. Let Americans disdain to be the instruments of European greatness! Let the thirteen states, bound together in a strict and indissoluble Union, concur in erecting one great American system, superior to the control of all transatlantic force or influence, and able to dictate the terms of the connection between the old and the new world!" For Hamilton, union, trade, and the theory of degeneracy were all intimately intertwined.

While Madison, Adams, Franklin, and Hamilton registered their thoughts on degeneracy in the New World, none was willing to challenge Buffon in any large-scale, coordinated manner. Jefferson, however, was both willing and able, and the case of American degeneracy was his to win or lose. Historian of science I. Bernard Cohen has argued that beside Theodore Roosevelt, Jefferson was the only American president "that could possibly be described as a would-be scientist."[15] Jefferson would need every ounce of his wits to take on Europe's premier scientific mind.

Thomas Jefferson was a naturalist at heart, and a man with a passion for science.[16] "Nature," Jefferson wrote a French colleague, "intended me for the tranquil pursuits of science, rendering them my supreme delight."[17] And Jefferson's claim that science was his passion while politics was his duty lends credence to the notion that had it not been for the extraordinary political circumstances leading to American independence—what Jefferson called "the boisterous ocean of political passions"[18]—he would have been content spending the majority of his time promoting science and losing himself in natural history, agriculture, the velocity of river currents, chemistry, archeology, anthropology, linguistics, meteorology, botany, the measurement of latitude and longitude, astronomy, and physics, to name just a few of his interests. Indeed, although he was never completely removed from politics, when he retired and was no longer actively holding office, Jefferson spent more and more time on natural history and gardening, and described himself as a "prisoner released from his chains."[19]

4.4. Thomas Jefferson made it one of his missions to debunk Buffon's theory of degeneracy."
Portrait by Gilbert Stuart.

The man who would one day send out Lewis and Clark to explore the
West had his interest in natural history kindled by early days roaming the
rich Piedmont forests that surrounded his boyhood homes at Shadwell
and Tuckahoe.[20] His father, Peter, a land developer and surveyor, taught
young Thomas about life in the back country, and instilled in his son a
passion for exploring unknown regions, which was reinforced by the
books in Peter's modest library—books such as *Maps of the 4 Quarters of
the World*.[21]

In 1752, nine-year-old Thomas was enrolled in Reverend William
Douglas's Latin School. He claimed to have learned little during the five
years he spent under Douglas, but things began to look up when, at age
fourteen, he entered Reverend James Maury's school. As an avid naturalist and collector himself, Maury rekindled Jefferson's interest in natural
history. Maury's fossils, minerals, and seashells were a center of discussion
both in school and at Maury's home, where Jefferson was a boarder.[22]

In 1760, Jefferson entered William and Mary College and quickly forged
a strong bond with William Small, an instructor who was only eight years

Jefferson's senior. Small was hired to teach mathematics, physics, and metaphysics, but as a result of the massive infighting between faculty, and the generally contentious relationships among those who ran William and Mary, he ended up teaching all the courses of philosophy offered in the curriculum. Jefferson was somewhat in awe of his new teacher and friend, viewing Small as "a man profound in most of the useful branches of science."[23] It was more than Small's new teaching method—one that involved interactive lecturing rather than mere rote assignments—or the eye-opening books, such as *A Course of Experimental Philosophy*, that drew Jefferson to Small. It was the long walks they took together that forged their relationship—walks in which Small would point out all the wonders around them, and teach his student how to "subject the physical world's phenomena to scientific analysis."[24]

When Jefferson entered William and Mary, he was a budding naturalist. When he left, largely as a result of Small and a few others,[25] he was enchanted with the scientific process, and would become one of its most vociferous promoters. In the years to come, he would speak of scientists "forming a great fraternity spreading over the whole earth. . . . Their correspondence is never interrupted by any civilized nation."[26]

A career in law did little to diminish Jefferson's passion for natural history, although the time he could devote to this subject was more limited. Politics would come to dominate much of his life, but he never stopped yearning for forays into the natural world. As secretary of state he longed to get away from the "detestable" drains on his time, so that he could go out and study the Hessian fly,[27] and as president he used one of the rooms in the newly built executive mansion to study fossils.[28] In his retirement, he decorated the entryway to Monticello with, among other things, his natural history collection. One visitor, who wrote about his time at Monticello in 1817, "supposed there is no private gentleman in the world, in possession of so perfect and complete a scientific, useful ornamental collection. To discuss them with his numerous visitors was one of Jefferson's pleasures in his later life."[29] Retirement was also a time devoted to establishing the University of Virginia, where Jefferson was keen on hiring "no one who is not of the first order of science in his line," even if that meant a little academic espionage that involved dispatching "a special agent to the Universities of Oxford, Cambridge & Edinburgh."[30]

Jefferson possessed what historian of science Silvio Bedini describes as a "compulsion to collect and record in pocket memorandum books random bits of information,"[31] many of which centered on natural history. He measured everything. No detail was too small for Jefferson to both note and enjoy—he told his daughter Patty, "there is not a sprig of grass that shoots uninteresting to me."[32]

Jefferson's passion for measurement—a passion that became quite useful when countering Buffon's claims in *Notes on the State of Virginia*—bordered on the obsessive. On the way to Paris to assume his ministerial position, he recorded daily the air temperature and the identity of birds and marine animals. He kept a "Garden Book" for almost fifty years, a "Weather Memorandum Book" from 1776 to 1820, and a "Farm Book" for matters dealing with agriculture. Jefferson would stop on the road to measure the diameter of a tree that caught his attention, or the size of a mule that interested him.[33] He measured his own activities, noting that he walked four miles and 264 yards per hour. As president, he would keep a schedule of when thirty-seven different vegetables appeared at market, and took notes on when this or that species of birds appeared, as well as when frogs started croaking.

Jefferson's obsession with measuring things relevant to natural history is also evident in the virtually endless list of things that Lewis and Clark were ordered to measure, including "the soil & face of the country, its growth & vegetable productions . . . the mineral productions of every kind; but more particularly metals, limestone, pit coal, & saltpeter; salines & mineral waters, noting the temperature of the last, & such circumstances as may indicate their character; volcanic appearances; . . . climate, as characterized by . . . the proportion of rainy, cloudy, & clear days, by lightning, hail, snow, ice, by the access & recess of frost, by the winds prevailing at different seasons, the dates at which particular plants put forth or lose their flower, or leaf, times of appearance of particular birds, reptiles or insects."[34]

In his later years, Jefferson would try to justify his obsession for collecting bits and pieces of data, writing that "a patient pursuit of facts, and cautious combination and comparison of them is the drudgery to which man is subjected by his Maker, if he wishes to attain sure knowledge."[35] But a justification hardly seemed necessary, as the man from the Piedmont forests reveled in the beauty and intricacies of the natural world.

It is not clear exactly when Thomas Jefferson first came across the theory of degeneracy, but we do know that in the early 1780s, he used his book, *Notes on the State of Virginia*, as a platform to voice his displeasure with the ideas of Buffon and Raynal.[36] Jefferson became embroiled in the question of degeneracy because he thought that Buffon's idea was conceptually unsound and that the data which Buffon relied upon to support his theory of degeneracy was flawed.

On the conceptual front, Jefferson saw no reason to believe that differences between the New World and the Old should translate into degeneracy in the former. His argument here is elegant and worth quoting at length:

> The opinion of a writer [Buffon], the most learned too of all others in the science of animal history, [is] that in the new world . . . nature is less active, less energetic on one side of the globe than she is on the other. As if both sides were not warmed by the same genial sun; as if a soil of the same chemical composition, was less capable of elaboration into animal nutriment; as if the fruits and grains from that soil and sun, yielded a less rich chyle, gave less extension to the solids and fluids of the body, or produced sooner in the cartilages, membranes, and fibres, that rigidity which restrains all further extension, and terminates animal growth. The truth is, that a Pigmy and a Patagonian, a Mouse and a Mammoth, derive their dimensions from the same nutritive juices.[37]

Jefferson did not deny that climate could affect the size of animals. Instead, he posited that that there was no evidence that differences existed in the climates of the two worlds that would lead one to think that life in the New World should be expected to degenerate compared to life in the Old World. "It is the uniform effect of one and the same cause, whether acting on this or that side of the globe," Jefferson proclaimed, and "it would be erring therefore against that rule of philosophy, which teaches us to ascribe like effects to like causes, should we impute this diminution of size in America to any imbecility or want of uniformity in the operations of nature."[38]

In a letter to his friend, the Marquis de Chastellux, Jefferson cuts to the heart of Buffon's argument that degeneracy was due to the cold and wet climate of the New World by citing America's premier thinker, and a man whom the French adored: "I am lately furnished with a fact by Dr. Franklin," wrote Jefferson, "which proves the air of London and of Paris to be more humid than that of Philadelphia, and so creates a suspicion that the opinion of the superior humidity of America may, perhaps, have been too hastily adopted. And supposing that fact admitted, I think the physical reasonings urged to shew, that in a moist country animals must be small, and that in a hot one they must be large, are not built on the basis of experiment."[39] While data on humidity were furnished by an impeccable source, Jefferson was always leery of generalizing from a single set of data, and he cautioned that the question of climate differences "cannot be decided, ultimately, at this day. More facts must be collected, and more time flow off, before the world will be ripe for decision. In the mean time, doubt is wisdom."[40]

Jefferson had a second conceptual bone to pick with Buffon; namely, that climate, even when it does affect the size of organisms, works within defined limits. For Jefferson, the upper and lower bounds of animal size was determined by a "Maker," who set "certain laws of extension at the time of their formation." What this meant was that even when climate could explain differences between groups of organisms, it was only so powerful a force: "All the manna of heaven," Jefferson writes, "would never raise the mouse to the bulk of a Mammoth."[41]

Above and beyond his conceptual arguments with Buffon—about what sort of differences one might expect between life in the Old and New Worlds—Jefferson felt obliged to attack the theory of degeneracy because of a more serious question regarding Buffon's data. Were they accurate, and if not, why? Were the data fraudulent, and if they were, who was to blame for such an egregious breach of civilized norms?[42]

These problems were especially troubling because they centered on Buffon—one of the leading natural historians in the world, and a man whom Jefferson admired. The Count, Jefferson noted, was a "celebrated Zoologist, who has added and is still adding, so many precious things to the treasures of science."[43] Buffon's theory of degeneracy, however, rested largely on the data of others for whom Jefferson did not have nearly the same regard.

Jefferson argued that Buffon's sources had "causally" collected their data and often based what they wrote on nothing more than hearsay. One such source was Peter Kalm, who was sent by the Swedish Academy to study the natural history of America. Though at times a very reliable source, Kalm was not always such.[44] On one occasion he wrote an incredible story in which he saw a bear kill a cow by biting into its hide. The bear then blew air into the cow's hide as hard as it could, and caused the cow to virtually explode.[45]

Jefferson's initial comments about Buffon's data seem harmless enough. "It does not appear," he noted, "that Messrs. de Buffon and D'Aubenton have measured, weighed, or seen [the wildlife] of America." The fact that Buffon and his colleague, both Frenchman in France, did not measure North American wildlife themselves is not all that surprising, as neither had been to North America. Jefferson then posed a critical question about the travelers upon whom Buffon relied: "Who were these travellers?" That is, were they reliable and trustworthy like Buffon and his colleagues working in the king's garden? "Was natural history the object of their travels?" Jefferson continued. "Did they measure or weigh the animals they speak of? Or did they not judge of them by sight, or perhaps even from report only? Were they acquainted with the animals of their own country, with which they undertake to compare them? Have they not been so ignorant as often to mistake the species?"[46] These were serious questions, if not accusations, and Jefferson was well aware of that. But they were merited, because a flawed theory about America, based on questionable data, was a very dangerous thing.

Although Jefferson had little data to judge the quality of Buffon's sources, he was skeptical of their objectivity. He was convinced that these sources had already made up their minds that the Old World was superior to the New in every way before they observed any animals in nature. Such men, of course, were not to be trusted on an issue as important as degeneracy. Indeed, Jefferson was convinced that if he could obtain honest answers to the myriad questions he posed, that "would probably lighten their authority, so as to render it insufficient for the foundation of an hypothesis."[47] Jefferson, though, was not sanguine about the prospects on this front, and bemoaned, "How unripe we yet are, for an accurate comparison of the animals of the two countries."[48]

As we'll see in more depth in the next chapter, Jefferson's problem with degeneracy and objective data gathering was not limited to animals. When it came to Buffon's, Raynal's, and de Pauw's claims about American Indians, Jefferson again found the data wanting, and this drew him deeper into the degeneracy debate. Although he knew little of South American Indians, and viewed Buffon's and the others' stories about them "to be just as true as the fables of Aesop,"[49] he rightly considered himself something of an expert on North American indigenous peoples. Nothing he saw in the European writings on American Indians struck him as accurate. He all but accused Buffon of sweet-talking his reader: "I am induced to suspect," Jefferson noted, "there has been more eloquence than sound reasoning displayed in support of this theory. . . . I must doubt whether in this instance he has not cherished error . . . , by lending her for a moment his vivid imagination and bewitching language."[50]

One can sense Jefferson's anger in response to de Pauw's renderings of American Indians, calling him "a compiler from the works of others; and of the most unlucky description; for he seems to have read the writings of travellers, only to collect and republish their lies." Worse still, Jefferson was convinced that de Pauw knew that the data that he was presenting were fictional. "It is really remarkable," Jefferson continued, "that in three volumes . . . of small print, it is scarcely possible to find one truth."[51] These were the strongest words that Jefferson ever used in his defense against the claim of degeneracy. Buffon may have used his glowing pen to seduce readers, but as Jefferson saw it, de Pauw was just an out-and-out liar. Franklin held a similar opinion of de Pauw, calling him "an evil-minded writer."[52]

Some of Jefferson's colleagues, like Benjamin Vaughn, were willing to take what today we would call a relativistic approach to degeneracy, and, in so doing, provided Buffon and his camp with an escape. In a 1787 letter to Jefferson, Vaughn wrote, "I have sometimes thought that if Eastern North America had been peopled by old and civilized nations some ages before the birth of Christ . . . [and they] had discovered Western Europe . . . similar remarks would probably have occurred (after a time) among *their* speculative philosophers, on the rude and ill provided condition of the New World."[53] But Jefferson would have none of that, and came to the conclusion that Buffon and his colleagues based their animal and human degeneracy arguments on a faulty conceptual base and (at best) suspect data.

Jefferson also had a practical reason for getting involved with the theory of degeneracy. He understood that the very survival of the United States of America rested in part on how its relations—especially trade relations—developed with European countries, and the extent to which people from other lands would immigrate to the United States. These issues were especially salient to Jefferson when he served as American minister to France from August 1784 through October 1789,[54] since his primary mission in Paris was to "negotiate treaties of commerce with foreign nations in conjunction with Mr. Adams and Dr. Franklin."[55] Though the ideas on degeneracy began in France, Jefferson's overarching concern was not the United States' relation with that country or its people. Jefferson understood France to be the only country "on which we can rely for support, under any event," and the French people "love us more, I think, than do any other nation on earth."[56] It was the pernicious effects of degeneracy on America's relationship with the rest of Europe that worried Jefferson.

Commercial treaties already existed between the United States and France, Sweden, and the Netherlands when Jefferson arrived in Paris in 1784, but no such treaties were in place with other European countries, nor with their colonial holdings in the West Indies.[57] Indeed, although he was generally pessimistic about the prospects, in a February 1785 letter to James Monroe, Jefferson spoke of the establishment of commercial trade with Britain, Spain, Portugal, Denmark, and numerous other European countries, arguing that such relationships were an "important part of our business."[58]

Yet, if Buffon and his colleagues were correct, why should any country develop trade relations with America, a degenerate land?[59] As Gilbert Chinard, who specialized in Jefferson's views on nature, noted, "If America proved unable to develop a large population, if the climate was not normally healthy and the soil normally productive . . . the American experience would fail to fulfill the hopes and expectations of the European liberals. These questions, which at first were purely philosophical or speculative, became, after 1776, political problems of vital importance."[60]

Moving from the sphere of geopolitics to that of the individual, Jefferson understood that if the idea of degeneracy remained unchallenged, then why would anyone emigrate to the United States, forcing his family to endure the long haul from Europe to America? Indeed, we have already seen that Frederick the Great had a strong antiemigration policy when it came to America, and that de Pauw may have been trying to ingratiate himself to the king by writing his *Philosophical Researches*. Jefferson thought this sort of problem could be general, and that it might affect the very survival of the United States.

Jefferson had already seen the way that natural history in the form of the degeneracy argument could be used as an economic and political tool against America. In 1777, shortly after the Revolutionary War began, an Englishman named William Robertson authored a famous book entitled *History of America*,[61] in which he argued that "the principle of life seems to have been less active and vigourous there," and that the climate had "proved pernicious to such as have migrated into it voluntarily . . . or have been transported thither by Europeans."[62] No wonder," Robertson noted, "that the colonists sent from Europe were astonished at their first entrance" into the New World. "It appeared to them waste, solitary and uninviting."[63] The New World, then, was degenerate in just the way Buffon and others had claimed.[64] To make matters worse, Robertson's book was excerpted in many newspapers, including the *Pennsylvania Packet*, the *Massachusetts Spy*, and the *Continental Journal*,[65] which led to an even greater dissemination of the ideas he was promoting.

This was precisely the sort of propaganda that Jefferson feared was the natural outcome of Buffon's ideas—if they remained unchallenged. Degeneracy, which started out as a natural history argument, was being used as a tool against America. And for Jefferson, Robertson suffered from the same sourcing problem that plagued Buffon, only doubly so: "As to Robertson," Jefferson wrote, "he never was in America, he relates nothing on his own knowledge, he is a compiler only of the relations of others, and a mere translator of the opinions of Monsr. de Buffon."[66]

Historian and political scientist Philippe Roger has argued that the theory of degeneracy created a scientific backbone to claims of European superiority.[67] Jefferson understood this all too well—and without the benefit of hindsight. Because of this he was willing to devote considerable time and effort into challenging the world's premier naturalist. And he would issue his challenge in the most direct way possible in *Notes on the State of Virginia*—by providing his own natural history data to debunk degeneracy. For Jefferson, this was the only respectable route to take. If Buffon's data was based on natural history, then so must be his response. Indeed, Jefferson himself would come to view natural history as a tool to express American pride, imploring Joseph Willard, the president of Harvard University, that "the Botany of America is far from being exhausted, its Mineralogy is untouched, and its Natural History or Zoology, totally mistaken and misrepresented." The study of these subjects was necessary "to do justice to our country, its productions and its genius. It is the work to which the young men, whom you are forming, should lay their hands. We have spent the prime of our lives in procuring them the precious blessing of liberty. Let them spend theirs in shewing that it is the great parent of science and of virtue."[68]

CHAPTER FIVE

"Geniuses Which Adorn
the Present Age"

The fact that Thomas Jefferson had the time to write *Notes on the State of Virginia,* with its harsh critique of degeneracy, is remarkable. By all accounts, and by any reasonable standard of what to expect from even the most brilliant of men, he should not have had that time.

On June 1, 1779, in the midst of the American Revolution, thirty-six-year-old Thomas Jefferson was elected governor of Virginia. Honored at his selection, within a month he was bemoaning the responsibilities of public office, and spoke of "the hour of private retirement [as] . . . the most welcome of my life."[1] As governor, Jefferson's concerns for the commonwealth included procuring money, recruiting soldiers, and supplying those soldiers with weapons and food. Though these were serious and difficult matters, for the first year and a half of Jefferson's gubernatorial term, Virginians had been spared the ignominious fate that had befallen other states like New York and Massachusetts—invasion by British troops. Then, in late fall of 1780, Benedict Arnold and his invasion force crossed into Virginia.

In May 1781, under threat of attack by British forces, Jefferson left the capital in Richmond and rode home to Monticello.[2] His departure has often been portrayed as the desperate act of a frightened man, but Jeffersonian scholar Dumas Malone has argued otherwise: "This was not a precipitate flight, but a dignified, if discouraged, retirement before a new alignment of superior forces."[3] Still, it was an extraordinarily difficult period for Jefferson, and would have been even more so, if an unexpected opportunity to immerse himself in natural history had not arisen.

That opportunity arose in the form of a series of queries posed by the Marquis de Barbé-Marbois, a man whom New Hampshire governor John Sullivan referred to as "one of those usefull genuiss [*sic*] that is constantly in search of knowledge."[4] In the early 1780s, Marbois—who more than a decade later would negotiate the Louisiana Purchase with then President Jefferson—was the secretary of the French legation in the United States. One of his assignments was to gather information on the fledgling states of the union, and so he designed a series of probing questions that he circulated around a number of states, including Virginia, Delaware, New Jersey, and New Hampshire.[5]

Some time in the fall of 1780, Marbois had completed his list of twenty-two queries about Virginia, which he presented to Joseph Jones.[6] Jones, an uncle of James Monroe and a member of the Continental Congress, passed the questions on to Governor Jefferson. Initially, Marbois was unaware that his queries had been forwarded to Jefferson,[7] but by the late winter of 1781 that had changed, and the two men were exchanging letters on that very subject.

Many of Marbois' twenty-two queries centered on topics related to natural history. He requested "some details" relating to "an exact description of its [Virginia's] limits and boundaries"; "the history of the State"; "a notice of the counties, cities, townships . . . caverns, mountains, productions, trees, plants, fruits, and other natural riches"; "a notice of the best sea ports"; "notice of the mines and subterranean riches"; "some samples of these mines and their extraordinary stones. In short, a notice of all what can increase the progress of human knowledge."[8]

No man in Virginia was better suited to field Marbois' queries than Jefferson. His breadth of knowledge, his love for Virginia, and his passion for both history and natural history made him the ideal choice as respondent. Looking back at the occasion in later years, Jefferson recalled, "I had always made it a practice whenever an opportunity occurred of obtaining any information of our country, which might be of use to me in any station public or private, to commit it to writing. These memoranda were on loose papers, bundled up without order, and difficult of recurrence when I had occasion for a particular one. I thought this a good occasion to embody their substance, which I did in the order of Mr. Marbois' queries."[9] And equally important, Jefferson possessed a "canine appetite" to learn more about everything,[10] and would use Marbois' questions as an excuse to immerse himself in this new mission. He quickly realized that working on answering these queries would make him much more acquainted with natural history and Virginia, and that he could put the information he gleaned to good use.

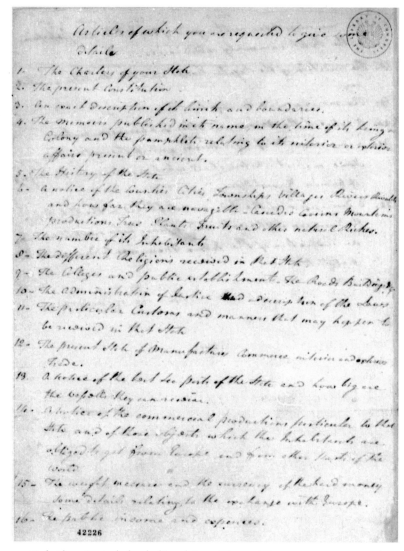

5.1. Marbois' queries, in the hand of Joseph Jones. Library of Congress.

Given that the queries reached Jefferson at the same time that the British were invading Virginia, it is hardly surprising that he initially requested some time to gather and organize his thoughts, informing Marbois that "my present occupations disable me from compleating" a response.[11] Over time, Jefferson became apologetic about the delay: "I fear your patience has been exhausted in attending them," he wrote Marbois, "but I beg you be assured there has been no avoidable delay on my part."[12]

Jefferson finally sent Marbois a reply to his queries on December 20, 1781—this reply, in slightly modified form, would eventually be published in book form as *Notes on the State of Virginia*.[13] In particular, Marbois had asked about the "productions, trees, plants, fruits, and other natural riches" of Virginia.[14] This section, which evolved into Query VI in *Notes on the State of Virginia* was where Jefferson took the opportunity to go well beyond the borders of Virginia, and discuss natural history as it related to animals in all the United States. Query VI was by far the longest and most detailed section of *Notes*. It was also where Jefferson directly attacked Buffon's theory of American degeneracy.

It would take three and a half years for Jefferson's reply to Marbois to be printed in book form as *Notes on the State of Virginia*, and much has been written about this transformation. Initially, Jefferson circulated his answers to Marbois' questions to just a few colleagues, including the adventurer and explorer George Rogers Clark.[15] Soon, others learned about this work and asked for copies. The length of Jefferson's reply to Marbois made additional handwritten copies for these new solicitors "too laborious,"[16] and Jefferson decided to look into a publisher for *Notes*.

In 1784, while serving as a member of the Continental Congress in Philadelphia, Jefferson discussed *Notes* with a publisher named Robert Aitken, but was quoted a price that he felt "exceeded the importance of the object."[17] In addition, Jefferson was unhappy with the time it would take Aitken to run off copies of the book.[18] As a result, he took a manuscript form of *Notes* with him when he boarded the *Ceres* and sailed to France as a new minister plenipotentiary.[19] Once in Paris, he planned to print the volume in France, provided, of course, that he could find an English printer in Paris.

Historians first believed that Jefferson employed a Parisian printer by the name of Monsieur La Marche to produce a few hundred copies of *Notes*. In his account book, Jefferson had made note of paying La Marche "for sheets," which was a common term used for printed pages in the late eighteenth century. It turns out that La Marche sold linens, and the sheets Jefferson had made reference to were house linens.[20]

Jefferson eventually found his Paris printer in Philippe-Denis Pierres. In May 1785, Pierres printed two hundred copies of *Notes*, with Jefferson's name intentionally left off the work, and the year 1782 (rather than 1785) listed on the cover page. These copies were meant for Jefferson alone to distribute, not for "the public at large."[21] He beseeched the few friends who had a copy to "peruse it carefully," and keep it to themselves.[22] Once he thought that the distribution of the book was under his control, Jefferson immediately saw to it that a copy was delivered to Buffon via their mutual colleague, the Marquis de Chastellux.[23]

Jefferson's attempts at restricting *Notes* to the eyes of his friends were in vain. Word of the book spread quickly, and he was overwhelmed with requests for copies. To make matters worse, Jefferson learned that a Monsieur Barrios had surreptitiously obtained a copy and was going to translate it into French without permission. Jefferson thought he had found a solution to this problem in the person of the Abbé Morellet, a member of

N O T E S

ON THE

STATE OF VIRGINIA.

WRITTEN BY

THOMAS JEFFERSON.

ILLUSTRATED WITH

A MAP, including the States of VIRGINIA, MARY-
LAND, DELAWARE and PENNSYLVANIA.

L O N D O N:

PRINTED FOR JOHN STOCKDALE, OPPOSITE
BURLINGTON-HOUSE, PICCADILLY.

M.DCC.LXXXVII.

5.2. Table of contents of Jefferson's *Notes on the State of Virginia*. In 1785, Parisian printer
Philippe-Denis Pierres printed two hundred copies of *Notes*, with Jefferson's name
intentionally left off the work, and the year 1782 (rather than 1785) listed on the cover page.
(This figure is actually of the 1787 Stockdale edition.)

the Académie française. The plan was for Morellet to translate the book
with Jefferson's help, thereby circumventing an unauthorized edition by
Barrios. Unfortunately, Jefferson's collaboration with Morellet did not go
well, and he described the Morellet French edition that appeared in early
1787 as "so wretched an attempt at translation . . . abridged, mutilated,
and often reversing the sense of the original, I found it a blotch of errors
from beginning to end."[24] With such an assessment, it is not surprising
that the front page of the book lists only Jefferson's initials.[25]

Fearful that the public would think that the French version of *Notes*
was accurate, in July 1787, Jefferson arranged for the first approved-for-
sale-to-the-public English edition of *Notes* to be published by British
bookseller, John Stockdale. A year later, an American edition of *Notes*
was published by Prichard and Hall (Philadelphia).[26] Jefferson had made

some minor revisions to his work since the original replies to Marbois, but aside from a beautiful introductory map and a bit of double-checking on sources he used to counter Buffon's degeneracy argument in Query VI, the British and American editions are similar to the manuscript Jefferson had with him when he sailed to France in 1784.[27]

Notes was generally well received—it was used as a popular handbook for natural history and geography and reprinted in newspapers all over the United States.[28] Over the centuries *Notes* has become mythologized. Historians speak of it in glowing terms, describing Jefferson's book as "probably the most important scientific and political book written by an American before 1785," "one of the first masterpieces of American literature," "one of America's first permanent literary and intellectual landmarks," and "perhaps, the most frequently reprinted book ever written in the United States south of Mason and Dixon's line."[29]

Jefferson was never fully satisfied with *Notes*. Perhaps it was the perfectionist in him, but at times he would speak of his book as "bad," and of himself as "the author of which has no other merit than that of thinking as little of it as any man in the world can."[30] In subsequent years, Jefferson tinkered with the idea of revising and updating the book. As late as 1809, the sixty-year-old ex-president was still telling his friends, "The Notes on Virginia, I have always intended to revise and enlarge, and have, from time to time, laid by materials for that purpose."[31] He did not abandon the idea completely until 1814, when he wrote that a revision of *Notes* "was no more to be entertained."[32]

In *Notes*, Jefferson worked through the first five queries posed by Marbois (regarding the boundaries of Virginia, rivers, seaports, mountains, and cascades) in fairly short order. These sections of his book were not meant as a rebuke to Buffon's degeneracy theory per se, but they do leave the reader with a sense of awe for all things Virginian. Jefferson described his state as "one third larger than the islands of Great Britain and Ireland."[33] The mountains were described in majestic language, with Jefferson writing that "the scene is worth a voyage across the Atlantic." Virginia's "natural bridge" of stone left observers speechless: "It is impossible for the emotions arising from the sublime, to be felt beyond what they are here," he wrote, "so beautiful an arch, so elevated, so light, and springing as it were up to heaven, the rapture of the spectator is really indescribable!"[34]

Query VI was where Jefferson tackled Buffon's theory about American life, but the first section of this query centered on minerals, and contained nothing per se about degeneracy. But Jefferson did use this opening section of Query VI to begin his quarreling with Buffon. During the course of a discussion about "an immense amount of *schist*" rock near one of

Virginia's mountains, Jefferson wrote of coming across shells "of very different kinds . . . to any I have ever seen in the tide-waters." These shells, along with some others found on the mountaintops in the Andes, Jefferson told his reader, have been taken by some people "as proof of a universal deluge."[35]

Jefferson discussed three theories for the great flood—one of which centered on Buffon's idea that the sea was once a lake, and a second in which Voltaire suggested that some shells were not produced by animals at all—only to dismiss all these ideas for lack of evidence. "The three hypotheses are equally unsatisfactory," Jefferson demurred, "and we must be contented to acknowledge, that this great phaenomenon is as yet unsolved. Ignorance is preferable to error; and he is less remote from the truth who believes nothing, then he who believes what is wrong."[36] This section ably demonstrates Jefferson's general reliance on empirical evidence to settle disputes regarding natural history and geology.

The next section of Query VI was on vegetables, and amounted to an accounting of things that grew in Virginia. And though Jefferson centered his argument against degeneracy on zoology, and not botany,[37] two points about this section are noteworthy. First, rather than a poisonous environment associated with a degenerate continent, Jefferson pointed out that Virginia boasted many *medicinal* plants. Second, following page after page of plants and their scientific names, Jefferson the epicurean tempted the reader by describing a Virginia where "the gardens yield musk melons, water melons, tomatas [*sic*], okra, pomegranates, figs, and the esculent plants of Europe. . . . The orchards produce apples, pears, cherries, quinces, peaches, nectarines, apricots, almonds, and plumbs [*sic*]."[38] Certainly not the sort of delicacies one would associate with a depauperate, degenerate environment.

Jefferson's counterattack on the degeneracy argument began in earnest in the section of Query VI called "Animals." This section opened with a rather long and odd description of a debate between Buffon and Jefferson over whether a huge "mammoth" once roamed North America[39] (a debate to which we'll return). After his sortie into mammoths, Jefferson outlined for the reader of *Notes* what Buffon had claimed about American degeneracy: "The opinion advanced by the Count de Buffon, is 1. That the animals common both to the old and new world, are smaller in the latter. 2. That those peculiar to the new, are on a smaller scale. 3. That those which have been domesticated in both, have degenerated in America: and 4. That on the whole it exhibits fewer species."[40]

Jefferson took on each claim in order, but rather than compare the entire New and Old Worlds, he chose to focus on a representative of each:

America for the New World, and Europe for the Old. Readers of *Notes* would have been curious as to why the world's most famous naturalist would make such sweeping claims of American inferiority, and so Jefferson summarized Buffon's reasons: that "the heats of America are less; that more waters are spread over its surface by nature, and fewer of these drained off by the hand of man. In other words, that heat is friendly, and moisture adverse to the production and developement of large quadrupeds."[41]

While Jefferson had reason to suspect that America was not more humid than parts of the Old World,[42] in *Notes,* he asked his reader to give the benefit of the doubt to Buffon, and to suppose that America *was* more humid. Even were that true, Jefferson noted, for animal life to thrive, plants were needed, and plants required both moisture and heat. The New World's humidity, then, was one of the two most important factors required for producing an environment *healthy* for animal development. A comparison of Europe and America, with respect to animal life, produced a draw: "Let us take two portions of the earth, Europe and America for instance, sufficiently extensive to give operation to general causes," Jefferson argued,

> let us consider the circumstances peculiar to each, and observe their effect on animal nature. America, running through the torrid as well as temperate zone, has more heat, collectively taken, than Europe. But Europe . . . is the dryest. They are equally adapted then to animal productions; each being endowed with one of those causes which befriend animal growth, and with one which opposes it.[43]

Jefferson's comparison of Europe and America nicely demonstrates that he was comfortable using *Notes* to reach well beyond the borders of Virginia to make his arguments against degeneracy. His comparison of heat and humidity across America and Europe also highlights the fact that Jefferson was not trying to show that New World animals were superior to those in the Old World, but rather to demonstrate that Buffon's claim about the superiority of the Old World was incorrect, both in principle and in fact.

Eighteenth-century natural historians had a passion for numbers, particularly numbers associated with size,[44] and they would often summarize their findings in tables of one sort or another. Jefferson was no different, and he relied on a systematic use of tables to counter Buffon's claims, one by one. With respect to animals, Jefferson listed twenty-five such species (plus the mammoth) that were common to both Europe and America, and despite his worries regarding the reliability of some of Buffon's sources,

| A Comparative View of the Quadrupeds of Europe and of America. |||
| *I. Aboriginals of both.* |||
	Europe	America
	lb.	lb.
Mammoth		
Buffalo. Bison		*1800
White bear. Ours blanc		
Caribou. Renne		
Bear. Ours	153.7	*410
Elk. Elan. Orignal, palmated		
Red deer. Cerf	288.8	*273
Fallow deer. Daim	167.8	
Wolf. Loup	69.8	
Roe, Chevreuil	56.7	
Glutton. Glouton. Carcajour		
Wild cat. Chat sauvage		†30
Lynx. Loup cervier	25.	
Beaver. Castor	18.5	*45
Badger. Blaireau	13.6	
Red Fox. Renard	13.5	
Grey Fox. Isatis		
Otter. Loutre	8.9	†12
Monax. Marmotte	6.5	
Vison. Fouine	2.8	
Hedgehog. Herisson	2.2	
Martin. Marte	1.9	†6
	oz.	
Water rat. Rat d'eau	7.5	
Wesel. Belette	2.2	oz.
Flying squirrel. Polatouche	2.2	†4
Shrew mouse. Musaraigne	1.	

5.3. Jefferson's tables 1–3 from *Notes on the State of Virginia*.

the data that he presented on European species was based on the French naturalist's figures. The weights of the American species were, in Jefferson's eyes, provided by "judicious persons, well acquainted with the species."[45] The American species on the table were represented by either the actual weights of specimens "deemed among the largest of their species"[46] (denoted by a ★ in Jefferson's table) or estimates of the weights of the larger specimens that such judicious people had seen in nature (denoted by a † in Jefferson's table). The weight of many American entries, however, were left blank, as Jefferson simply did not have the relevant data.

Jefferson's table I showed the weights for six species that lived in both America and Europe, with the American animals larger in five cases. In the sixth case—the red deer—the weights were close enough that Jefferson argued that they were essentially equal. For many of the blank entries in this table, Jefferson discussed anecdotes about the relative size of the species in question. From the weights and the anecdotes, he concluded that seven species were about the same size in Europe and America, seven were larger in America, and the other twelve cases were indeterminate. Rather than suggest American superiority, Jefferson took a more conservative

II. Aboriginals of one only.			
Europe		**America**	
	lb.		**lb.**
Sanglier. Wild boar	280.	Tapir	534.
Mouflon. Wild sheep	56.	Elk, round horned	†450.
Bouquetin. Wild goat		Puma	
Lievre. Hare[69]	7.6	Jaguar	218.
Lapin. Rabbet	3.4	Cabiai	109.
Putois. Polecat	3.3	Tamanoir	109.
Genette	3.1	Tamandua	65.4
Desman. Muskrat	oz.	Cougar of N. Amer.	75.
Ecureuil. Squirrel	12.	Cougar of S. Amer.	59.4
Hermine. Ermin	8.2	Ocelot	
Rat. Rat	7.5	Pecari	46.3
Loirs	3.1	Jaguaret	43.6
Lerot. Dormouse	1.8	Alco	
Taupe. Mole	1.2	Lama	
Hamster	.9	Paco	
Zisel		Paca	32.7
Leming		Serval	
Souris. Mouse	.6	Sloth. Unau	27¼
		Saricovienne	
		Kincajou	
		Tatou Kabassou	21.8
		Urson. Urchin	
		Raccoon. Raton	16.5
		Coati	
		Coendou	16.3
		Sloth. Aï	13.
		Sapajou Ouarini	
		Sapajou Coaita	9.8
		Tatou Encubert	
		Tatou Apar	
		Tatou Cachica	7.
		Little Coendou	6.5
		Opossum. Sarigue	
		Tapeti	
		Margay	
		Crabier	
		Agouti	4.2

approach: "The first table impeaches the first member of the assertion, that of the animals common to both countries, the American are small-est."[47]

Jefferson used table II to counter Buffon's assertion that there were more uniquely European animal species than American species, and that animals in Europe were larger. In this table he listed eighteen species unique to Europe, but a whopping seventy-four species that were found only in America. What's more, the average weight of the eighteen European species was approximately twenty-seven pounds, while the American animals weighed in at about an average of fifty-five pounds. Together, these findings on species number and size led Jefferson to conclude that the "second table disproves the second member of the assertion, that the animals peculiar to the new world are on a smaller scale, so far as that assertion relied on European animals for support: and it is in full opposition to the theory which makes the animal volume to depend on the circumstances of heat and moisture."[48]

II. Aboriginals of one only.

Europe	America	
	Sapajou Saï	3.5
	Tatou Cirquinçon	
	Tatou Tatouate	3.3
	Mouffette Squash	
	Mouffette Chinche	
	Mouffette Conepate. Scunk	
	Mouffette. Zorilla	
	Whabus. Hare. Rabbet	
	Aperea	
	Akouchi	
	Ondatra. Muskrat	
	Pilori	
	Great grey squirrel	†2.7
	Fox squirrel of Virginia	†2.625
	Surikate	2.
	Mink	†2.
	Sapajou. Sajou	1.8
	Indian pig. Cochon d'Inde	1.6
	Sapajou. Saïmiri	1.5
	Phalanger	
	Coquallin	
	Lesser grey squirrel	†1.5
	Black squirrel	†1.5
	Red squirrel	10. oz.
	Sagoin Saki	
	Sagoin Pinche	
	Sagoin Tamarin	oz.
	Sagoin Ouistiti	4.4
	Sagoin Marikine	
	Sagoin Mico	
	Cayopollin	
	Fourmillier	
	Marmose	
	Sarigue of Cayenne	
	Tucan	
	Red mole	oz.
	Ground squirrel	4.

III. Domesticated in both.

	Europe	America
	lb.	lb.
Cow	763.	*2500
Horse		*1366
Ass		
Hog		*1200
Sheep		*125
Goat		*80
Dog	67.6	
Cat	7.	

When it came to Buffon's third claim—that domesticated animals imported from the Old World degenerate in America—Jefferson's table III listed eight species. But this table was essentially useless, both because it contained so few species, and in only one case (the cow) did it list weights in both Europe and North America. Instead of relying on his table, Jefferson conceded that in many cases, North American animals "have become less than their original stock,"[49] but the reason had nothing to do with Buffon's degenerative forces (cold and moisture), and everything to do with the vast natural resources found in the United States, combined with the country's small human population.

Jefferson argued that domesticated animals in America were on the paltry side only because farmers let them graze freely on the vast tracts of land and forest that surrounded them: the farmer "finds it more convenient to receive them from the hand of nature . . . than to keep up their size by a care and nourishment which would cost him much labour."[50] If European farmers chose the same approach, their animals would look like those found in America. Agricultural and farming strategies, not degeneracy, explained why some domesticated animals were not as hardy in America as Europe. This was more than an isolated agrarian argument by Jefferson, as it had philosophical implications as well: "It would be erring therefore against that rule of philosophy, which teaches us to ascribe like effects to like causes, should we impute this diminution of size in America to any imbecility or want of uniformity in the operations of nature."[51] Overall, Jefferson viewed Buffon's third assertion regarding degeneracy, "as probably wrong as the first and second were certainly so."[52]

With respect to New World degeneracy and animals, that left but one claim; namely, that "on the whole it [America] exhibits fewer species." Here, Jefferson posited that all his tables "taken altogether" refuted Buffon's last claim. Further evidence against Buffon's contention was added in the form of a detailed list of one hundred and seventy "Birds of Virginia" tacked on to the end of Query VI. This table wasn't comparative in the way that the others were—that is, the birds of Virginia were not meant to be put side by side with the birds of some European locality to address the question of degeneracy. Still, the list was clearly an attempt by Jefferson to provide his readers with additional fodder against claims about America's inferior wildlife. Jefferson was also hinting to his readers that Buffon should have known better in the first place, as information on the one hundred and seventy species of birds found in Virginia came, in part, from Buffon's *Natural History*.

Jefferson's tone was low-key for most of Query VI. This changed dramatically when he addressed Buffon's claims about degeneracy in Native

Americans—here his writing took on an air of both passion and competitiveness.[53] "I can speak of him somewhat from my own knowledge," Jefferson wrote of the North American Indian in Query VI, "but more from the information of others better acquainted with him, and on whose truth and judgment I can rely."[54] From this accumulated knowledge base, Jefferson systemically countered many of the claims made by Buffon:

[The Indian] is neither more defective in ardor, nor more impotent with his female, than the white reduced to the same diet and exercise . . . he is brave, when an enterprize depends on bravery; education with him making the point of honor consist in the destruction of an enemy by stratagem, and in the preservation of his own person free from injury; or perhaps this is nature; while it is education which teaches us to honor force more than finesse . . . he will defend himself against an host of enemies, always chusing to be killed, rather than to surrender, though it be to the whites, who he knows will treat him well: that in other situations also he meets death with more deliberation, and endures tortures with a firmness unknown almost to religious enthusiasm with us. . . . He is affectionate to his children, careful of them, and indulgent in the extreme. . . . His affections comprehend his other connections, weakening, as with us, from circle to circle, as they recede from the center. . . . His friendships are strong and faithful to the uttermost extremity. . . . His sensibility is keen, even the warriors weeping most bitterly on the loss of their children, though in general they endeavour to appear superior to human events.[55]

Although Jefferson spoke of friendship and vivacity of the mind, the above quotation does not directly address whether Indians were degenerate in terms of overall mental abilities. And though he believed "more facts are wanting," Jefferson was fairly certain that in the long run, "we shall probably find that they are formed in mind as well as in body, on the same module with the 'Homo sapiens Europaeus.'"[56]

To further convince the readers how misguided the idea of Indian degeneracy was, Jefferson wanted to present examples of "eminence in oratory," but argued that such examples were hard to find because Indian speeches were "displayed chiefly in their own councils" and hence inaccessible to most whites.[57] There were, however, exceptions, such as a speech made by Chief Logan.

Jefferson began his discussion of Logan with the sort of glowing praise that was typically reserved for whites during the early days of the republic: "I may challenge the whole orations of Demosthenes and Cicero, and of any more eminent orator, if Europe has furnished more eminent, to produce a single passage, superior to the speech of Logan, a Mingo chief, to Lord Dunmore, when governor of this state."[58] Given Jefferson's love of all things from ancient Greece, it is hard to imagine a more powerful introduction to Logan.

Logan's story began in the spring of 1774, when two Shawnee tribesmen were involved in the robbery and murder of "inhabitants of the frontiers of Virginia."[59] In response, a Colonel Cresap, whom Jefferson described as "a man infamous for the many murders he had committed on those much-injured people," pulled together a group to avenge the settler's death.[60] "Cresap and his party concealed themselves on the bank of the river," Jefferson wrote, "and the moment the canoe reached the shore, singled out their objects, and, at one fire, killed every person in it."

The canoe was filled with Logan's family members. Before that moment Chief Logan had been "a friend of the whites,"[61] but soon after the killings, a war ensued between the Virginia militia and Logan's tribe.[62] The militia defeated the Indians, and in despair Logan sent Lord Dunmore, the governor of Virginia, the words that so moved Jefferson:

> I appeal to any white man to say, if ever he entered Logan's cabin hungry, and he gave him not meat; if ever he came cold and naked, and he clothed him not. During the course of the last long and bloody war, Logan remained idle in his cabin, an advocate for peace. Such was my love for the whites, that my countrymen pointed as they passed, and said, "Logan is the friend of white men." . . . Col. Cresap, the last spring, in cold blood, and unprovoked, murdered all the relations of Logan, not sparing even my women and children. There runs not a drop of my blood in the veins of any living creature. This called on me for revenge. I have sought it: I have killed many: I have fully glutted my vengeance. . . . Logan never felt fear. He will not turn on his heel to save his life. Who is there to mourn for Logan?—Not one.[63]

Jefferson immediately followed the transcription of Logan's speech by noting that it was all the more remarkable because the alphabet had not yet been introduced to Native Americans.[64]

Jefferson's contention was that degeneracy didn't explain any differences between whites and Indians.[65] "I do not mean to deny that there are varieties in the race of man . . . both of their powers of body and mind," he told his readers. "I only mean to suggest a doubt" whether such differences "depend on the side of the Atlantic on which their food happens to grow, or which furnishes the elements of which they are compounded? Whether nature has enlisted herself as a Cis or Trans-Atlantic partisan?"[66] Jefferson thought not, and was certain that the reason that others disagreed was Count Buffon's "vivid imagination and bewitching language."[67]

Jefferson devoted more space to his spirited defense of North American Indians than to any other of Buffon's claims.[68] To the modern reader, his passion on this issue begs the question—what of degeneracy and blacks?[69] Were blacks inferior, and was there evidence that any differences between whites and blacks were a function of the American environment?

Jefferson's thoughts on the relation between whites and blacks were complex—here we shall focus only on what he said about whites, blacks, and degeneracy in *Notes on the State of Virginia*. Perhaps the most salient point about degeneracy is that the topic is never mentioned in Jefferson's response to Query VI. No mention of blacks is made in reference to the theory of degeneracy in Jefferson's direct response to Buffon; not one.[70] Jefferson's discussion of blacks and of slavery are, instead, embedded in his Query VIII ("The Number of Its Inhabitants") and Query XIV ("The Administration of Justice and Description of Laws").

One reason that Jefferson spoke of Indians but not of blacks in Query VI was that although Buffon the anthropologist had many condescending things to say about African people throughout *Natural History*, his long and vicious attack on Native Americans was specifically part of his claims on American *degeneracy*, and so Jefferson tailored his response in kind. A more general answer to why Jefferson omitted blacks from his discussion of degeneracy per se has yet to emerge. One possibility, suggested by historian Dumas Malone, was that the relationship between Indians and whites was a philosophical issue for Jefferson, and because philosophy and natural history were tightly linked in the eighteenth century, he discussed Indians in the context of natural history (Query VI).[71] But the relationship between blacks and whites was a practical, legal, and economic one for Jefferson, and so he discussed it elsewhere in *Notes*.[72] When these relations were discussed in *Notes*, they caused a firestorm.

At the very end of his discussion of population size in Query VIII, Jefferson made a remark that more than any other was responsible for his initial reticence to publish *Notes* in a public forum. In the context of a prior Virginia bill outlawing the importation of slaves, Jefferson wrote, "This will in some measure stop the increase of this great political and moral evil, while the minds of our citizens may be ripening for a complete emancipation of human nature."[73] Jefferson realized that calling slavery a moral evil was a dangerous stand for a Southern statesman, and that his call for gradual emancipation would not be well received by many: he wanted to publish these ideas only when the scales were tipped in his favor with respect to "whether their publication would do most harm or good,"[74] as he had no wish "to be exposed to . . . censure."[75]

Censure followed nevertheless, and, throughout his political career,

Jefferson's opponents pointed to his slavery comments in *Notes* to paint Jefferson as a danger to the economy of the South and the nation.[76] At one point, he was so distraught about the way people were using his comments on Indians, blacks, and slavery in *Notes* that Jefferson lamented, "'O! that mine enemy would write a book!' has been a well known prayer against an enemy. I had written a book, & it has furnished matter for abuse for want of something better."[77]

Jefferson brought up the emancipation of black slaves again in Query XIV (Administration of Justice and Description of Laws). There, in the context of a bill to be brought before the Virginia Assembly, he outlined a plan to emancipate all slaves born after the bill was passed, and to train them "at public expense to tillage, arts and sciences." Once that was accomplished, blacks were to be sent out and "colonized to such place as the circumstances of the time should render most proper, sending them out with arms, implements of household and of the handicraft arts, feeds, pairs of the useful domestic animals . . . to declare them a free and independent people, and extend to them our alliance and protection."[78]

In *Notes,* Jefferson argued that colonization of blacks was required because of "deep rooted prejudices entertained by the whites; ten thousand recollections, by the blacks, of the injuries they have sustained."[79] What's more, colonization was inevitable because of the "real distinctions which nature has made" between blacks and whites.[80] These differences, Jefferson wrote, included skin color as well as assorted physiological distinctions—blacks "secrete less by the kidneys, and more by the glands of the skin, which gives them a very strong and disagreeable odour. This greater degree of transpiration renders them more tolerant of heat, and less so of cold, than the whites."[81] But nowhere does Jefferson ever suggest that the difference between the races was somehow associated with Buffon's supposed differences between the environments of the Old and New Worlds.

The reader of *Notes* senses a Jefferson who is unsure about the status of blacks relative to whites. Are blacks inferior, even if it has nothing to do with Buffon's ideas on degeneracy? Over and over, Jefferson seems to qualify his answer to this question in one way or another. As he continues to elaborate on the "real" differences between the races, Jefferson's uncertainties emerge again and again. Blacks are "at least as brave, and more adventuresome" than whites, *but* "this may perhaps proceed from a want of forethought, which prevents their seeing a danger till it be present." When it comes to love, "They are more ardent after their female," *but* Jefferson adds, that love "seems with them to be more an eager desire, than a tender delicate mixture of sentiment and sensation."[82]

In terms of the cognitive processes of blacks as compared to whites, "in memory they [blacks] are equal to the whites." On the other hand, when it came to the ability to invoke reason, blacks were "much inferior," Jefferson wrote, "as I think one could scarcely be found capable of tracing and comprehending the investigations of Euclid; and that in imagination they are dull, tasteless, and anomalous."[83] There were no black poets, painters, or sculptors that Jefferson knew of—even among blacks that had been "liberally educated." When it came to music, however, blacks "are more generally gifted than the whites with accurate ears for tune and time." Even here though, Jefferson backtracks, adding "Whether they will be equal to the composition of a more extensive run of melody, or of complicated harmony, is yet to be proved."[84]

Jefferson saw inherent differences between blacks and whites,[85] but told the reader of *Notes* that "to justify a general conclusion, requires many observations." "I advance it therefore as a suspicion only," he wrote of his theory about whites and blacks. "Blacks," he continued, "whether originally a distinct race, or made distinct by time and circumstances, are inferior to the whites in the endowments both of body and mind."[86] And so we are left with a Jefferson who, in *Notes,* supported emancipation, but believed that blacks were inferior. This inferiority was not, however, mentioned in the context of degeneracy. Indeed, in *Notes,* while it is clear that Jefferson did not claim to understand the causes of the supposed inferiority of blacks, nowhere does he suggest that this inferiority had anything to with the cold and moist environment that was so central to Buffon's degeneracy hypothesis.[87]

Degeneracy was Buffon's brainchild, and, for the most part, Jefferson directed his comments in Query VI toward the French naturalist. In *Notes,* Jefferson ignored de Pauw, but could not bring himself to do the same when it came to one of Raynal's claims about degeneracy—that America had not produced a single good poet, a mathematician, or a genius in any field. With respect to poets, Jefferson did not deny Raynal's claim, he simply argued that it was unreasonable, and that America needed time to groom such talent: "When we shall have existed as a people as long as the Greeks did before they produced a Homer, the Romans a Virgil, the French a Racine and Voltaire, the English a Shakespeare and Milton," Jefferson noted, "should this reproach be still true, we will enquire from what unfriendly causes it has proceeded."[88]

Jefferson was not so conciliatory about Raynal's other claims. Surely, he argued, Raynal should have known better than to claim that America had not produced a world-class man of science—what of Franklin, with whom Raynal had dined, and who was a major celebrity in Raynal's own country? Such an omission was unconscionable for Jefferson, for "no one of the present age has made more important discoveries, nor has enriched philosophy with more, or more ingenious solutions of the phaenomena of nature" than Benjamin Franklin.[89] Jefferson's defense of American science was not restricted to Franklin, as he reminded his reader that "Mr. Rittenhouse [is] second to no astronomer living."[90]

Raynal's sweeping claim about the lack of genius also allowed Jefferson to point out America's contribution to the art and science of warfare. In this case it was personal for Jefferson, as he saw General Washington as a man "whose memory will be adored while liberty shall have votaries, whose name will triumph over time, and will in future ages assume its just station among the most celebrated worthies of the world." Washington alone proved Raynal wrong, Jefferson argued, and it was absurd that Raynal would group the general "among the degeneracies of nature."[91] What's more, these examples of genius were the product of a young, underpopulated America, proving Raynal's claim to be "as unjust as it is unkind. When it came to the "geniuses which adorn the present age," Jefferson pronounced, "America contributes its full share."[92]

In his response to Query VI, Jefferson used the opportunity to counter the degeneracy argument with all the data he could muster, and with the prosaic dexterity that today's reader associates with his pen. But he wasn't finished yet—for he still hoped to hand Buffon a present that might, if he were lucky, put the degeneracy argument to rest once and for all.

CHAPTER SIX

Enter the Moose

Though *Notes on the State of Virginia* was his only book, Jefferson's writings—especially his correspondence—were prolific, and show him to be the quintessential man of ideas and thought. But Jefferson also understood that in order to convince, you sometimes need to go with the tangible—things that people can hold, touch, feel, and smell. This approach would serve him well in the argument over degeneracy.

While serving in France, Minister Plenipotentiary Jefferson planned to personally hand the great naturalist Buffon proof that American life was not inferior to that in the Old World—and that proof would come in the form of a seven-foot-tall moose. Jefferson could be only hesitantly optimistic that this strategy would work. Twice before, Buffon had been given physical evidence that his images of life in the New World were incorrect. His responses on those occasions did not bode well for Jefferson's plan to use the giant moose to persuade the Count to recant his degeneracy hypothesis.

The first pre-moose incident occurred just before Jefferson was to sail off to his ministerial post in France. Jefferson had long suspected that Buffon had mislabeled the American panther as the cougar in *Natural History*. Buffon had claimed that the panther was found in Africa and the East Indies and possessed "a ferocious air, a restless eye, a cruel aspect, brisk movements . . . has a rough and very red tongue, strong and pointed teeth, hard sharp claws, a beautiful skin."[1] Cougars, on the other hand, which Buffon initially thought were restricted to South America, but then learned were in North America (as black cougars), were less impressive: "their skin is so tender as to be easily pierced by the simple arrows of the Indians. . . . Neither the jaguars nor cougars are absolutely ferocious. . . . They despise the assaults of dogs, whom they often seize in the very neighbourhood of houses. When pursued by such a number of dogs as obliges them to fly, they take refuge in the trees."[2]

Jefferson was certain that the cougar that Buffon had identified in America was in fact a black panther, as panthers so often killed sheep and wreaked general havoc that bounties were offered for them and reported in newspapers such as the *New York Journal*, the *Boston Gazette*, and the *Charleston Morning Post*.[3] An opportunity for Jefferson to get his hands on evidence that he could use to make the case that the panther was indeed found in America arose in May 1784, when Jefferson and his daughter Patsy stopped in Philadelphia en route to boarding a France-bound ship from Boston.[4] As Jefferson recalled the story: "I observed an uncommonly large panther skin, at the door of a hatter's shop. I bought it for half a Jo on the spot, determining to carry it to France to convince Mons. Buffon of his mistake with relation to this animal: which he had confounded with the Cougar."[5] It would take some time, however, before Jefferson would be able to present both his argument and his panther skin to Buffon. In October 1785, more than year after arriving in France, Jefferson wrote that he had "never yet seen Monsr. de Buffon,"[6] which is not all that surprising given that Jefferson was stationed in Paris, while Buffon spent a great deal of time at his family estate back in Montbard.

At last, in the first few days of 1786, Jefferson got his chance to meet the Count in person, when Buffon offered to host a dinner for Jefferson and the Marquis de Chastellux.[7] Jefferson had sent the panther skin to Buffon before the dinner—indeed receipt of the panther skin seems to have spurred the dinner invitation—and he talked with Buffon about this creature during his dinner visit. He remembered Buffon as "a man of extraordinary power in conversation,"[8] and at the dinner, Jefferson presented his case for the panther. The Count was gracious, and after listening to Jefferson's arguments, "he acknowledged his mistake."[9] At the same time, Buffon gave no indication of revising his general assessment of North American life, let alone his ideas on degeneracy.

On the surface, Jefferson's wish that Buffon admit his error about the panther seems picayune—more a display of Jefferson's interest in the minutiae of taxonomy than an issue relevant to the argument about North American degeneracy. It is true that the Count painted a more noble picture of the panther than of the cougar, but there is no evidence that Jefferson thought the panther vs. cougar argument was especially relevant to Buffon's claims about degeneracy. What this episode demonstrated, though, was that Jefferson would go a long way to prove Buffon wrong. Voyages across the Atlantic were lengthy and arduous, and people were careful to take with them only their most valued possessions. Yet in the tumult of preparing to go from Philadelphia to Paris, Jefferson thought it important enough to take time to stop at a hatter, buy a panther skin, and carry it on board—all to make a point with Buffon.

Having Buffon admit his error was so significant an event that Jefferson was still telling the story to Daniel Webster more than forty years later.[10] The whole affair was an attempt to demonstrate to Buffon, and others who would one day read of it in Jefferson's correspondence, that the Count didn't know his North American animals particularly well. And if he was wrong about panthers and cougars, then his credibility was all the lower when it came to major issues like American degeneracy.

The second pre-moose instance—wherein Jefferson encountered in Buffon a man who seemed to refuse to budge, even in the face of physical evidence—revolved around the "mammoth" discussed in *Notes on the State of Virginia*.

In the summer of 1705, a Dutch farmer unearthed a huge tooth weighing nearly five pounds on the grounds of Claverack Manor, near Albany, New York.[11] The tooth was sold to Albany assemblyman Peter van Bruggen for a half a pint of rum, displayed before the assembly, and eventually given to the royal governor of New York, who shipped it off to the Royal Society in a box marked "tooth of a giant."[12]

More bone fragments of this giant creature were soon found around the original site, and the news spread. Congregationalist minister Edward Taylor, whose grandson Ezra Stiles would become president of Yale University and an important figure during the Revolutionary War period, had seen the bones for himself, and he believed that they were the remains of a "monster" that he judged to be above "60 or 70 feet high."[13] Taylor even composed a poem about the so-called "giant of Claverack:"

His Arms like limbs of trees twenty foot long,
Fingers with bones like horse shanks and as strong
His thighs do stand like two vast millposts stout.[14]

Soon the American colonists, as well as others interested in these remains, were engaged in the first debate about the Claverack monster— were these the remains of a race of giant humans, or some giant animals?[15] While some prominent figures, such as Cotton Mather, held that the bones were those of the giants of the Old Testament—giants that had perished in Noah's flood—most believed them to be the remains of some terrible animal akin to that described in Taylor's verse. This argument continued for many years, but the point became moot in 1728 after a pair of Royal Society papers that were based on comparative anatomy demonstrated that the remains were clearly those of some giant animal—an animal that would play a role in the American degeneracy argument.[16]

Interest in the remains of this giant Claverack creature grew when news that prisoners who had been exiled to Siberia by Peter the Great had uncovered similar fossil remains thousands of miles away. The Russians spoke of these remains as those of a "mammut" or mammoth.[17] Soon the colonists were using this term to describe their findings as well, although many referred to the mysterious creature whose bones had been uncovered as the incognitum—a term derived from the Latin for "unknown."[18]

More and more bones of the incognitum were uncovered in the years following. Perhaps the most important of these were the 1739 remains uncovered at a salt lick—called the "Big Bone Lick"—in what would eventually become the Commonwealth of Kentucky. This salt-lick site became the major source of mammoth bones in the colonies—Benjamin Franklin had four ivory tusks, a molar, and three vertebrae from the site[19]—with bones being unearthed there well past the American Revolution. So many fossil remains were uncovered at what soon became known as "Big Bone,"[20] that a sample of mammoth bones was eventually shipped to Paris—bones that became the focus of another argument between Buffon and Jefferson.

As the number of fossil remains grew, so too did the legend of the incognitum. Its sheer size, and the very notion that something that big roamed North America, led many colonists to believe that this was a fierce creature, and that image carried immense power. Eventually the mammoth was painted as a killer of enormous proportion, capable of apocalyptic actions: "Forests were laid waste at a meal, the groans of expiring animals were every where heard," read one description, "and whole villages, inhabited by men, were destroyed in a moment."[21] Although that depiction did not appear until 1801, Thomas Jefferson understood the power that the incognitum represented twenty years earlier, both when writing *Notes on the State of Virginia,* and during his time in Paris. Such a creature could play a role in debunking the degeneracy claim. If Buffon accepted that the mammoth—the largest terrestrial creature ever described at the time—had roamed North America, how could its existence be squared with his degeneracy argument?

The bones from the Big Bone Lick that were sent to Paris became part of Buffon's Royal Cabinet of Natural History, and the Count himself examined these samples. He first mentioned both the American and Siberian samples in a rather unexpected place—toward the end of his 1761 description of lions and tigers in volume 9 of *Natural History*. There Buffon wrote of his "astonishment" at these creatures that dwarfed the elephant, but which "no longer exist." Buffon's views—both of a new species, distinct from the elephant, and of the extinction of this species—would vacillate over time. But, at that moment, Buffon argued not only that the mammoth was extinct, but that extinction might be common: "How many smaller, weaker, and less remarkable species must likewise have perished, without leaving any evidence of their past existence?"[22]

Buffon's collaborator, the world-famous anatomist Louis-Jean-Marie Daubenton, examined the remains of the mammoth sometime after the Count's reference to its extinction. By comparing the Big Bone Lick bones (as well as the bones of the Siberian mammoth) with those of the elephant, he found no significant differences, and concluded that the mammoth was nothing more than a large—albeit very large—elephant. Daubenton, however, recognized one flaw in his theory. While the bones matched those of the elephant, the teeth from the lick did not neatly match an elephant's "grinders." He concluded that the teeth (grinders) were those of a hippopotamus that had died in the same area, and were incorrectly lumped in with the elephant bones. Such a mistake would be easy for Daubenton to imagine, as it was well known that many of the bones at Big Bone Lick were collected by Indians, and the theory of degeneracy would suggest that with such savages doing the collecting, errors were bound to occur.

Buffon next took up Daubenton's position and made it explicit in his 1764 *Natural History* (volume 11) entry on the elephant. From Daubenton's observations, Buffon noted, "we cannot doubt that those tusks and bones we have already noticed for their prodigious size, actually belonged to the elephant. . . . Daubenton is the first who has proved them unquestionable by exact measures and comparisons, and reasons founded on the great knowledge that he has acquired in the science of anatomy."[23] So while Buffon made it clear that America presently had no elephants, he accepted that they had once roamed that land. Then, in a second turnabout, in 1778, Buffon used his *Epochs of Nature* to return to his original position on extinction. New evidence on jaw and tooth structure led Buffon to claim that the American and Siberian bones represented two different species, distinct from the elephant, and both extinct. He named the American species the American mastodon. Indeed, George Gaylord Simpson, one of the twentieth century's leading experts on fossils, has argued that the whole debate around the bones of the mammoth and mastodon marked the birth of some of the tenets of modern paleontological thinking.[24]

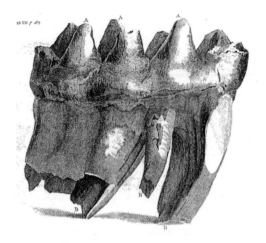

6.1. Fossil evidence from the "grinders" of the American incognitum led Buffon to ultimately see the incognitum as a distinct species (and much larger) than the elephant. Buffon's final position was that both species went extinct.

For Buffon, the argument about the bones of the mammoth and the mastodon was not really about degeneracy, but rather about the role of extinction. But to Thomas Jefferson, those same bones were not only about extinction, but also very much about the degeneracy debate.

Jefferson examined parts of the incognitum himself and sought to expand his collection from a tooth and a thigh bone.[25] His correspondences with explorer George Rogers Clark (older brother to William Clark, of Lewis and Clark) are filled with discussions of the "big bones" that Jefferson was pestering his good friend Clark to collect and send. Jefferson described such bones as a "most desirable object . . . and there is no expense of package or of safe transportation that I will not gladly reimburse to procure them safely."[26]

Jefferson discussed the mammoth—he never used the term "mastodon"—in *Notes on the State of Virginia,* but his stance on this creature appears strange until we realize that he probably did not know about Buffon's final flip-flop on extinction in *Epochs of Nature.*[27] In *Notes,* then, Jefferson was attacking Buffon's first (1761) and second (1764) positions on the incognitum. These positions infuriated Jefferson for two reasons. Buffon's first stand involved the idea of extinction, and Jefferson believed that extinction was impossible. Even though Buffon pulled back from extinction after Daubenton's examination of the bones, Jefferson was angry that Buffon ever mentioned it to begin with. Part of Buffon's second position—that the bones were just those of the elephant—rankled Jefferson because of the implications for degeneracy. The incognitum was gigantic, and just the sort of thing Jefferson could use to fight the degeneracy claim (recall that he probably did not know that Buffon ultimately agreed with him on the size and distinctness of the mammoth).

Jefferson discussed the mammoth at length in his attack on Buffon's degeneracy theory. Indeed, in his table "A Comparative View of the Quadrupeds of Europe and of America," the mammoth is the very first animal listed. For Jefferson, the evidence for the mammoth as a distinct, very large new species was indisputable: "It is well known that on the Ohio, and in many parts of America further north, tusks, grinders, and skeletons of unparalleled magnitude, are found in great numbers, some lying on the surface of the earth, and some a little below it."[28] In addition to the hard physical evidence for the mammoth, there were the Indian tales of this great beast that seemed to have transfixed the future president.

In his introduction to this creature, Jefferson tells his readers, "Our quadrupeds have been mostly described by Linnaeus and Mons. de Buffon. Of these the Mammoth, or big buffalo, as called by the Indians, must certainly have been the largest."[29] Jefferson goes into more depth about the Indian legends of the mammoth, quoting a Delaware Indian chief as saying that "in an[c]ient times a herd of these tremendous animals came to the Big-bone licks, and began an universal destruction of the bear, deer, elks, buffaloes, and other animals."[30] Between the Indian tales, his examination of the bones, and the sheer magnitude of the physical evidence, Jefferson was convinced that the mammoth was a real North American animal, and that it was enormous.

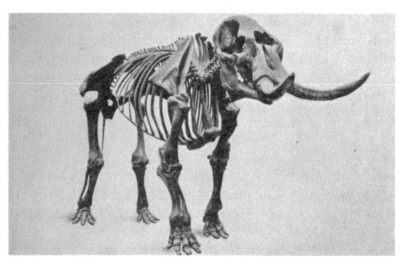

6.2. A reconstructed mastodon skeleton known as the Cohoes mastodon. Peale's mastodon first appeared in the halls of the American Philosophical Society.

Jefferson found it remarkable that European naturalists were so naive about the true nature of this animal, and he soon laid out what he thought was a deathblow to Buffon's second position that the bones were just those of an elephant mixed with those of a hippopotamus:

> It is acknowledged, that the tusks and skeletons are much larger than those of the elephant, and the grinders many times greater than those of the hippopotamus, and essentially different in form. Wherever these grinders are found, there also we find the tusks and skeleton; but no skeleton of the hippopotamus nor grinders of the elephant. It will not be said that the hippopotamus and elephant came always to the same spot, the former to deposit his grinders, and the latter his tusks and skeleton. For what became of the parts not deposited there? We must agree then that these remains belong to each other, that they are of one and the same animal."[31]

And if that weren't enough, Jefferson told the reader of *Notes* that the argument that the bones actually were those of an elephant did not fit with what was known about elephants, as they would never be able to survive in the more temperate climate of North America. "I have never heard an instance," he wrote, "and suppose there has been none, of the grinder of an elephant being found in America. . . . The elephant is a native only of the torrid zone and its vicinities."[32] Page after page of *Notes* continues this line of reasoning. Jefferson saw no escape for Buffon and his theory of North American degeneracy: "It is certain such a one has existed in America, and that it has been the largest of all terrestrial beings," he began in one of his most direct thrusts on degeneracy. Once that was ascertained, "It should have sufficed to have rescued the earth it inhabited . . . to have stifled, in its birth, the opinion of a writer, the most learned too of all others in the science of animal history, that in the new world . . . nature is less active, less energetic on one side of the globe than she is on the other."[33] But it didn't suffice in the slightest.

For Jefferson, the mammoth was not only a distinct species and very large, but it still roamed his country. He clung to the American Indian myths that the creature "still exists in the northern parts of America."[34] To a large extent, Jefferson's thought here stemmed from his view that extinction was incommensurate with Nature's laws. "Such is the oeconomy of nature," he told the readers of *Notes*, "that no instance can be produced of her having permitted any one race of her animals to become extinct; of her having formed any link in her great work so weak as to be broken."[35] Jefferson's Deist views on God and nature wouldn't allow for extinction. While he couldn't be sure that Buffon's discussion of extinction made the Count an atheist (although one suspects Jefferson may have thought that), Buffon was clearly taking a secular position on extinction. And the Count's view—which implied that nature was not the handiwork of God—did not sit well at all with Jefferson.

To allow for the mammoth's continued existence somewhere in the northern plains of America, Jefferson had to resort to tortured logic: "It may be asked, why I insert the Mammoth, as if it still existed? I ask in return, why I should omit it, as if it did not exist?" Many parts of America, he continued, "still remain in their aboriginal state, unexplored and undisturbed by us, or by others for us. He may as well exist there now, as he did formerly where we find his bones."[36] In other words, since we have no evidence regarding places we have not been, we should accept this as positive evidence that things that once existed still exist in these unexplored regions—they have just not yet been found.

Despite his attempts to use the incognitum as a feather in his cap, Jefferson knew that this creature could never get him what he really wanted—an admission from Buffon that North American life was not degenerate. While he thought that the odd mammoth might still be roaming in some undiscovered region of North America, he had nothing close to evidence for that. To demonstrate that North America was not degenerate in real time, Jefferson would need to hand Buffon physical evidence that *contemporary* life in North America was incommensurate with the idea of degeneracy. Enter the moose.

Jefferson spent a good deal of time compiling the numbers that went into his size comparison charts (Europe vs. America) that played such an important role in *Notes on the State of Virginia*. He knew that as a mathematician, Buffon would pay attention to his statistics. But Jefferson also understood that if he could present Buffon with some dramatic contemporary example that showed that the degeneracy theory was out-and-out wrong, that would have an equally important, if not more important, effect on the Count.[37]

A very large moose—preferably one that was about seven feet tall—would serve Jefferson's purpose. At the time, that would have seemed like a reasonable enough goal: Peter Kalm had written that the moose was at least as tall as the horse;[38] and at about the same time Jefferson was fixated on disproving degeneracy, Ezra Stiles, naturalist and president of Yale, wrote of "a Moose Calf" that was already close to three feet tall—and that was a female.[39] Surely, a full-grown male moose would meet Jefferson's expectations. And so Jefferson promptly set about trying to find such a creature. He began the process by gathering as much information as he could about the moose, while at the same time beginning what would prove to be a long search for just the right specimen to present to Buffon.

Jefferson's interest in procuring a good-sized moose,[40] or at least parts of one, is apparent from early on in his *Notes on the State of Virginia* project. In the same 1782 letter in which he asked George Rogers Clark for mammoth remains from Big Bone Lick, Jefferson made a plea for "horns of very extraordinary size," noting that such horns "would be very acceptable."[41] About a year and a half later, Jefferson began sending out a sixteen-question survey on the habits, size, and natural history of the moose to his colleagues, with explicit instructions for the survey information to be gathered from the most reliable hunters and woodsmen. In addition to the survey questions, Jefferson made it clear that if these hunters and woodsmen could procure for him the skeleton of a giant moose, he would be deeply indebted.

Jefferson's moose survey inquired about such natural history facts as "Do they make a loud ratling sound when they run?" to "What is their Food?" as well as practical issues such as "Have they ever been tamed and used to any purpose?"[42]

The answers to such questions would help Jefferson put together a more complete picture of the natural history of the moose, and would strengthen his position when he confronted Buffon. But for the purpose of presenting Buffon with a specimen that would help discredit the degeneracy theory, the heart of this survey lay in question 3: "What is the height of the Grey Moose . . . its length from ear to ear . . . its circumference where largest?"[43]

One of these surveys was sent to William Whipple in January 1784.[44] Whipple, who was a signer of the Declaration, as well as a general in the Revolutionary War, understood the importance of the moose to Jefferson. Whipple began his reply of March 15, 1784, with an apology for taking so long to respond to Jefferson's letter, and excused the delay by noting that he was still "expecting more satisfactory information on the moose." Unable to obtain all the information he wished, but knowing that Jefferson wanted answers as soon as possible, Whipple presented two sets of survey answers—his own and that of an unnamed gentleman "who never measured a moose tho' he has seen many."

The unknown gentleman listed the moose as four-and-a-half to five-and-a-half feet tall "when full grown." Whipple also spoke of his own experience with a one-year-old female moose, which he "judged to be four and a half feet high but was not measured when I saw it." Despite Whipple sending along a "small parcel of the hair of a moose," and a remark that he *heard* some hunters note that moose may grow to six feet tall, his reply must have disappointed Jefferson, who had his sights set on a much larger animal to help him debunk Buffon's degeneracy hypothesis. What's more, Whipple wasn't able to send a skeleton, even of the four-and-a-half-foot-tall moose. Buffon would need solid proof—bones—before Jefferson could convince him of anything, and Whipple showed no signs of procuring such concrete evidence.

Things were somewhat more promising when it came to general John Sullivan's response to Jefferson's moose queries. Knowing the moose to be common to New Hampshire, Jefferson had spoken with Sullivan in Annapolis, presented him with the natural history survey, and made his desire for a moose skeleton very clear.[45]

Sullivan, who was a representative at the Second Continental Congress and had been a prisoner during the battle for independence, was attorney general of New Hampshire at the time that Jefferson approached him about the moose for Buffon.[46] In his reply to Jefferson's natural history questions on March 12, 1784, Sullivan informed him that he had sent the survey out to many individuals, but that only one person—a Colonel McDuffee—had replied. Sullivan, however, did note that he was soon expecting more information from a man named Jonathan Dore, who had "been taken by the Indians when an infant, and remained with them thirty years [and] became one of them and has hunted with them in every part of America north of the Ohio."[47]

McDuffee reported that he had seen many moose that were even larger than Jefferson had imagined, with "one of the largest [being] about 8 1/2 feet high." Sullivan added a postscript, noting, "There is a moose's horn in one of the Pigwackett towns so large that it is used for a cradle to rock the children in." That said, no mention was made yet of any moose skeleton that Jefferson could actually present to Buffon. But that would soon change, as Sullivan and his connections became the focal point of Jefferson's obsession with getting his hands on a giant moose.

General Sullivan's next letter, dated June 22, 1784, included two more survey responses—one from Jonathan Dore, and another from a hunter named Gilbert Warren.[48] Although Warren noted that a moose grew to about "18 hands," or six feet, Dore reported that the "moose when full grown is about ten feet high." News on actually securing the skeleton of an appropriately large stature was mixed. Sullivan told Jefferson that he "had procured from the head of a province of Main[e] a large pair of moose horns" which he would send along. It is unclear whether these antlers were in fact sent to Jefferson—probably not, as no future reference is made to them[49]—but in any case, a pair of antlers wouldn't suffice,

1. Is not the Caribou & the Black Moose one & the same Animal? No. The Caribou is between a Moose & a deer

2. Is not the Grey-moose & the Elk one & the same animal? & quite different from the former? They are very different animals

3. What is the height of the Moose? Eighteen hands

4. What is the difference between the black & the grey Moose? The black is the Male & grey the female of the same species
   Have they a solid or cloven hoof? Cloven

5. Do their feet make a loud rattling as they run? They do

6. Is the under part of their hoof covered with hair? it is not

7. Are they a swift animal? Not very swift

8. Do they sweat when hard run or only drip at the tongue? They do not sweat

9. At what season do they shed their horns? in November And when recover them? in May

10. Has the Doe horns as well as the Buck? No

11. How many young does she produce at a time? Two

13. How far southward are they known? I know not

14. Have they ever been tamed & used for any purpose?
    They have been tamed but not used. I have seen a young one suck a cow as gently & kindly as her calf

6.3. Jefferson's queries on the moose, March 12, 1784 (as part of Sullivan's reply letter to Jefferson). Library of Congress.

15. Are the horns of the Elk palmated viz, flattened at
the top — or are they round & pointed? flattened at the top

16. Has the Elk always or ever a white spot a foot in
Diameter round the root of the Tail?
I never saw one

The above answers are from Gilbert Warren

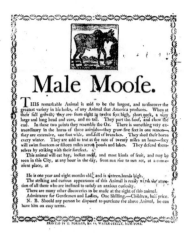

6. 4. A 1798 advertisement demonstrates the size and demeanor that early citizens of the United States attributed to the moose. The moose is described as "the largest . . . Animal that america produces, . . . from eight to twelve feet high." This broadside was printed in New York by G. Forman.

as Jefferson was intent on obtaining a full skeleton. Sullivan wrote that although he did not think he would be able to provide Jefferson with such a skeleton immediately, he was optimistic that he would be able to obtain such material by the following winter (the winter of 1784–1785).

Jefferson's passion—arguably his obsession—about the moose can be seen in his frequent exchange of letters on this subject with Sullivan. Sullivan told his friend that "the permission which your Excellency has given me of keeping up a correspondence affords me the highest pleasure and . . . a line from your Excellency will at all times be deemed an honor."[50] William Whipple had expressed similar sentiments in his correspondence with Jefferson about the moose, noting, "The pleasure I receive from this communication will be in proportion to the satisfaction it affords you."[51] It isn't clear whether Jefferson's interest in the moose was infectious, or whether Sullivan and Whipple were being deferential, but in any case, the hunt for a good giant moose skeleton was clearly on, and undertaken in earnest.

What makes Jefferson's moose-centered correspondence with Sullivan, Whipple, and others so remarkable is the context in which it occurred. His survey requests and pleas for a giant moose were sent out, and the replies came back, at the very time that Jefferson was corresponding with many of the other founding fathers over matters of national importance. For example, in the four-week span before Sullivan's letter of June 1784, Jefferson had exchanged letters with James Madison over disputes between various states in the fledgling republic, with George Washington regarding commercial treaties in Europe, and with John Adams and Benjamin Franklin regarding their appointments as ambassadors to France.[52]

Within two months of the above June 22, 1784, exchange with Sullivan, Jefferson was en route to Paris. There, absorbed with his new duties as minister plenipotentiary, and soon recognizing that the moose that Sullivan had promised to send had not arrived, Jefferson started to think that this particular avenue of attack would not come to fruition. Such thoughts vanished, however, after Jefferson dined with Count Buffon sometime in late 1785.

On January 7, 1786, shortly after his dinner with the Count, Jefferson wrote his friend Archibald Cary that Buffon was "absolutely unacquainted" with the American moose and deer.[53] Apparently Buffon thought that the moose was simply a miscategorized reindeer, which is much smaller than the moose.[54] In addition, Buffon had confused the American deer and the red deer of Europe.[55] "I attempted . . . to convince him of his error in relation to the common deer," wrote Jefferson, and "I told him that our deer had horns two feet long; he replied with warmth, that if I could produce a single specimen, with horns one foot long, he would give up the question. Upon this I wrote to Virginia for the horns of one of our deer, & obtained a very good specimen, four feet long."[56]

At that same dinner, in order to demonstrate how large North American animals truly were, Jefferson told Buffon "that the rein deer could walk under the belly of our moose." Buffon, Jefferson noted, "entirely scouted the idea."[57] The fact that Buffon refused to even consider the true size of the moose reinvigorated Jefferson, as he saw a new opportunity to show the Count the error of his ways.

Jefferson left the dinner meeting at Montbard with the distinct impression that Buffon would recant his theory of North American degeneracy—at least as it applied to animals—if he could show the Count evidence that the moose was a separate species, and one much larger than both the reindeer and the American deer. Buffon had given Jefferson every reason to think that if he were willing to "give up the question" of the size of the deer when presented with evidence of his error, then he would also "give up the question" of degeneracy when presented with even more dramatic evidence from the moose for the size of Northern American animals.

On the same day that he wrote Archibald Cary about Buffon's ignorance of the American moose, Jefferson immediately reissued his appeal for a moose from John Sullivan. With a new sense of urgency and a keener

Tom. III. *Pl. XVIII bis, pag .132.*

LA FEMELLE DU RENNE.

6.5. Buffon's rendition of the rein deer (from *Natural History*).

eye for detail, he beseeched Sullivan not to forget the quest for the moose
that they had discussed more than a year earlier:

> The readiness with which you undertook to endeavor to get for me the skin, the
> skeleton and the horns of the moose . . . emboldens me to renew my application
> to you for those objects, which would be an acquisition here, *more precious than you
> can imagine.* . . . I will pray you send me the skin, skeleton and horns just as you can
> get them. . . . Address them to me, to the care of the American Consul of the port
> in France to which they come.[58]

Jefferson then wrote a very similar request to William Whipple, again emphasizing that the moose skeleton "would be more precious than you could conceive."[59] And again, two weeks later, Jefferson penned a plea for a moose—this time to Archibald Stuart, who had served with him on a House of Burgesses committee back in Virginia. Jefferson implored Stuart to send him a moose or elk: "The most desireable form of receiving them would be to have the skin slit from the under paw along the belly to the tail, & down the thighs to the knee, to take the animal out, leaving the legs and hoofs, the bones of the head, & the horns attached to the skin by sewing up the belly & shipping the skin it would present the form of the animal."[60] Ever the epicurean, Jefferson tacked on a postscript: "I must add a prayer for some Paccan nuts, 100, if possible, to be packed in a box of sand and sent me."

In March 1786, Jefferson received a short reply from Sullivan in which the general noted that he still had the moose antlers he had mentioned in his correspondence of June 22, 1784. The letter, however, made no mention of obtaining an entire moose and it would be another nine months before he heard from Sullivan again.[61] Then, on January 26, 1787, the general wrote back with the most promising moose news so far.

After justifying his delay by noting the "want of opportunity," Sullivan told Jefferson that he had, through the hunters he worked with, secured a moose for his friend. The moose was not actually in Sullivan's hands just yet, but rather "on the Connecticut River," en route to Sullivan, and would arrive very soon. "I expect it in a sleigh as soon as the roads are broken through the snow which is now very deep," Sullivan told Jefferson, "and no time shall be lost in forwarding the same to your excellency."[62]

In a follow-up letter three months later, Sullivan informed Jefferson that he had received the moose, and proceeded to provide a more complete history of its capture. After receiving Jefferson's renewed request for a skeleton, Sullivan had contacted Captain Colburn to "procure me one [moose] and transport him to my house with the skins opened and the entrails taken out." Sullivan seems to have chosen his hunters well—even though the winter of 1787 "proved extraordinary," with "much snow,"[63] Colburn's group of twenty men killed a seven-foot-tall moose in Vermont.[64] It then took fourteen days "with a team in the transportation," before the moose arrived at Sullivan's house on April 3. Actually getting the moose to Paris would be another matter entirely.

The weeks of transport, in which Colburn and his team cleared a road twenty miles long,[65] had taken a heavy toll on the moose's carcass (not to mention Colburn's group), and once it was in Sullivan's hands, "every engine was set at work to preserve the bones and cleanse them from the remaining flesh. And to preserve the skins with hair on, with the hoofs on and bones of legs and thighs in skin without putrefaction." Unfortunately, Sullivan and his team had particular trouble preserving the skull bones and antlers. They reconstructed the moose as best they could, and Sullivan informed Jefferson that "the skin [of the head] being whole and well dresst it may be drawn on at pleasure." For some reason they were unable to save the antlers, but Sullivan sent another pair he had—"they are not the horns of this moose but may be fixed on at pleasure."[66]

The capture and preparation of the moose was more work than Sullivan ever imagined. "The job was both expensive and difficult," Sullivan informed Jefferson, and the entire ordeal was "a very troublesome affair."[67] Although Sullivan no doubt noted this to inform Jefferson that he was a gentleman of his word, another more practical reason was to justify the costs which Jefferson needed to reimburse. Sullivan noted that he had to borrow forty-five pounds sterling from his brother in order to ready the moose for transport.[68] Indeed, Sullivan sent a detailed "account of expenses" in which he broke down the cost of his endeavors, which included paying Captain Colburn for his time, securing the moose antlers, paying a tanner, securing a box big enough for shipping, shipping the box with the skeleton to port, and so on.[69] To be certain that Jefferson understood that this accounting had nothing to do with any profit motive, Sullivan reminded him, "I only charge for the expenses I have paid in cash, without any thing for my own trouble which has been very considerable."[70]

Sullivan's plan for sending the skeleton and skin of the seven-foot moose overseas—a plan he outlined to Jefferson in a series of letters that April (1787)—involved sending the remains from New Hampshire through England and finally to a port in le Havre-de-Grâce. Jefferson could then arrange pickup from this popular port of call in northern France. The transport was to be handled by Sullivan's colleague, captain Samuel Pierce. At first, the plan called for the moose remains to be transported out of New Hampshire on a ship that Pierce himself was taking to France, but Pierce then informed Sullivan he was not heading to France, but that he would take the box with him to England and would see to it that it was shipped safely to le Havre-de-Grâce.[71]

At this point in the story, the long time delay between sending and receiving trans-Atlantic mail started to bring about misunderstandings. Both on May 9, and again on May 29, 1787, Sullivan wrote to Jefferson that Captain Pierce's ship had left harbor at Portsmouth on the ninth of May. Unfortunately, the ship sailed without Sullivan's cargo for Jefferson, which was left at the port by Captain Pierce "either through accident or design."[72] In both these letters, Sullivan was quick to point out that he had contacted a Mr. De la Tombe, who was to arrange for the moose skin and skeleton to be shipped to le Havre-de-Grâce. Jefferson did not receive these communications until September,[73] and on August 15 he wrote Sullivan a depressing letter. Unaware of the events that had transpired regarding the moose, and having heard nothing more from Sullivan, Jefferson assumed that the remains had been lost for good. He also assumed that this marked the end to his quest for a moose to debunk the theory of degeneracy, and he wrote Sullivan, "Should the bones & skin of the moose been miscarried I would decline repeating the expense or giving your excellencey the trouble of a second sample."[74]

Even after Jefferson finally received Sullivan's letters in September, his spirits remained low. Just a few days after receiving Sullivan's second communication regarding the new shipping plans, Jefferson—who had yet to see the bones that were supposedly being shipped to him by Mr. De La Tombe—thought the whole moose affair over. In a September 28 letter to William Stephens Smith, Jefferson wrote that Sullivan had "put himself to an infinitude of trouble more than I meant, he did it cheerfully, and I feel myself really under obligations to him. That the tragedy might not want a proper catastrophe, the box, bones and all are lost; so that this chapter of natural history will still remain a blank. But I have written to him not to send me another."[75]

If Jefferson had waited just one or two more days, he could have saved himself much grief. Despite the pitfalls and delays, the remains of a seven-foot-tall moose arrived at le Havre-de-Grâce sometime between September 28—when Jefferson spoke of the catastrophe—and October 1. Immediately upon receiving the remains, an ecstatic Jefferson wrote to both Buffon and his anatomist colleague, Daubenton. Jefferson knew that Buffon would be at his Montbard country estate in October, so he sent the moose and one letter to Daubenton and asked that a second enclosed letter addressed to Buffon be forwarded to the Count.[76] Jefferson's letter to Buffon speaks of being "happy to be able to present you at this moment the bones and skins of a moose, [and] the horns of [another] individual."[77] Indeed, Jefferson hoped that the moose would be "stuffed and placed on his legs in the king's cabinet."[78]

Going into more detail about the specimen, Jefferson, somewhat apologetically, wrote to Buffon that "the skin of the moose was drest [sic] with the hair on, but a great deal of it has come off, and the rest of it is ready to drop off." Jefferson wanted Buffon to understand that he had completely miscategorized the moose as a much smaller reindeer, and informed the great naturalist that "the moose is perhaps a new class." A new class of beast, Jefferson hoped, that would help Buffon see the error of his ways regarding North American degeneracy.

Buffon apparently saw the moose that Jefferson had sent him, for Jefferson remembered that the skin and skeleton "convinced Mr. Buffon. He promised in his next volume to set these things right."[79] It is unclear exactly what Buffon was convinced of, or what he would set right, as we have no record of this from the Count. It seems reasonable to assume that Buffon was convinced that the moose and the reindeer were very different creatures, with the moose being much larger. Buffon may even have admitted that the moose did not fit in well with his theory of degeneracy. Whether he would have actually changed his mind on the whole degeneracy issue, as Jefferson had been led to believe he might, is hard to know. In any event, Jefferson's timing here was poor. No next updated volume with corrections would appear. Within six months of receiving Jefferson's moose, Count Buffon was dead.[80]

Jefferson believed that Buffon was not only *the* greatest natural historian of his generation, but a gentleman who was fair and objective. Therefore Jefferson could take comfort in the fact that the Count had seen such a stark example illustrating that life in America was anything but degenerate. That said, the effects of the moose on Buffon, whatever they might have been, would not be seen in any revisions to *Natural History*.

By early 1788, *Notes on the State of Virginia* was published in both French and English, Jefferson's moose was stuffed, and Buffon was dead. But the argument over New World degeneracy was anything but over.

CHAPTER SEVEN

Thirty-Seven-Pound Frogs and Patagonian Giants

On July 4, 1826, fifty years after signing the Declaration of Independence, Thomas Jefferson died. During his eulogy for the former president, New York senator Samuel Latham Mitchill, a man who himself mythologized the American mastodon to rebuke Buffon's ideas on American inferiority,[1] referred to Jefferson's campaign against degeneracy as the equivalent of proclaiming independence for a second time.[2] Mitchill himself was an ardent natural historian,[3] having founded the Lyceum of Natural History of New York City and served as professor of chemistry and natural history in the New York College of Physicians and Surgeons.[4] But Mitchill's comments give the impression that the degeneracy theory died at the hands of Jefferson. It did not; the idea that life in the New World was inferior survived Jefferson's passionate refutation. In Europe, it continued to be promulgated in one form or another by such men as Hegel, Keats, and Lamarck; European counterattacks on the degeneracy hypothesis would be launched by Philip Mazzei, Lord Byron, and Alexander von Humboldt.

Jefferson took a series of calculated risks in his point-by-point refutation of Buffon's degeneracy claim in *Notes on the State of Virginia*. He assumed that his analysis would convince even the most ardent skeptics that the Count was incorrect. Jefferson also gambled that if he countered Buffon's claims, he could largely brush aside Raynal and completely ignore de Pauw. This strategy made sense, as Buffon was not only the best known of the three, but his *Natural History* was the birthplace of the idea of degeneracy. And so Jefferson focused on Buffon, and devoted but a single—albeit long—paragraph retort to Raynal's book. And though he complained of it bitterly in letters to his friends,[5] in *Notes,* Jefferson never even mentioned de Pauw's scathing critique of all things American.

In these gambits, Jefferson underestimated the Eurocentric appeal of the degeneracy hypothesis—an appeal that lured thinkers from philosophy (de Pauw's book was among the few that the Marquis de Sade had in his prison cell at the Bastille),[6] from natural history, and from the arts and letters. Fortunately for Jefferson, there were European zealots on his side as well—people who saw the New World, and most especially the United States, as the fulcrum upon which the future rested. Indeed, one of the first counterattacks against the theory of degeneracy, written by a French Benedictine monk by the name of Antoine-Joseph Pernety, was published long before Jefferson penned *Notes.*

Within a year after the publication of de Pauw's book in 1768, Pernety had written a scathing reply entitled *Essay on America and the Americans against the Philosophical Researches by Mr. de Pauw.*[7] In addition to its appearing in print, Pernety read his thesis to the prestigious Academy of Berlin on September 7, 1769. The intelligentsia at the academy would hear Pernety use two very interesting cases against de Pauw's degeneracy argument.

In his book, de Pauw had argued that even though most American animals were smaller than those found in the Old World, there were a few exceptions to this rule. But, de Pauw claimed, these exceptions proved his point, for the few animals that flourished in the New World were vile— giant insects, venomous snakes, and the like that "so unhappily distinguish this hemisphere."[8] Added to this list of large, but inferior, creatures were the giant frogs of Louisiana that weighed in excess of thirty-seven pounds, and whose croaks sounded more like a young calf than a frog.[9]

In his reply to de Pauw, Pernety stood these giant frogs on their heads. Rather than deny de Pauw's claims, he embraced them. These *uber*-frogs were evidence that life in the New World was not degenerate. Pernety thought it absurd that de Pauw would tout Buffon's ideas for each and every instance in which Old World animals were larger than their New World counterparts, and at the same time claim that giant New World reptiles were evidence of degeneracy, just because reptiles were, in de Pauw's opinion, hideous, useless creatures. Pernety demanded consistency, but as he learned very quickly, that was not a priority of de Pauw's.

In his response to de Pauw, Pernety also made use of mythical Patagonian giants—a huge race of South American humans[10]—as evidence against the claim of degeneracy. The legend of giants in Patagonia dated back to 1520 when Ferdinand Magellan reached the southernmost part of

South America (modern day Argentina and Chile). Antonio Pigafetta, an Italian explorer who accompanied Magellan, and later wrote of his journeys in *Magellan's Voyages around the World*, told of a crewman who brought on board a Patagonian native who was so tall that the crew came up only to his waist. Rumors quickly grew, and the Patagonians became so large that they were said to "have eaten rats caught on the ship, whole, without even removing the skins!"[11]

The legend of the Patagonian giants spread and worked its way into popular culture in the late 1760s, when British admiral John Byron, commander of the HMS *Dolphin*, described encountering a Patagonian chief who was seven feet tall and noted that few of the others in the tribe were shorter.[12] This story was picked up not only in the European press, but by American periodicals such as the *American General Repository* and newspapers such as the *Connecticut Gazette,* which wrote of "between four or five hundred Patagonians of at least eight or nine feet in height."[13] Pernety took all this quite seriously: how could de Pauw argue for New World degeneracy in the face of a race of New World giants?

In 1771, de Pauw responded to Pernety in a book entitled *Defense of the Philosophical Researches on the Americans.*[14] Here de Pauw continued where he had left off in his original book, in which he had confidently proclaimed that no such Patagonian giants had ever existed, for if they had, surely "some living proofs of their existence would certainly have been brought to Europe: or, at least, their skeletons."[15] Though willing to publish every slanderous remark he could obtain on life in the New World without the slightest hint of evidence, de Pauw found the Patagonian giant story too incredible to entertain, and spent thirty pages in his rejoinder to Pernety arguing against such a race of South Americans. The same de Pauw that wrote of Indian tribes in which mothers molded the heads of their infants into the shapes of cones and squares[16] was unwilling to believe equally outrageous tales that didn't fit in with his philosophical position on New World inferiority.[17] Moreover, de Pauw was completely unapologetic about this.

It is tempting to view the Pernety–de Pauw exchange over such things as giant men, oversized frogs, and cone-headed Indian babies as evidence of two men with too much time on their hands. But for both Pernety and de Pauw, the theory of degeneracy was the issue of their day, and they took it very seriously. Their exchange provides the first evidence of how passionate the debate was within Europe, and the fact that de Pauw could not so safely be ignored.[18] In this instance, Pernety the European stood up to defend the New World against de Pauw and others—but this was a very rare occurrence early in the history of the degeneracy debate.

7.1. Lord John Byron encounters a Patagonian giant.

De Pauw's original book was published in French in 1768, and for the next two decades, the English-speaking world came into contact with his ideas only secondhand, through the works of such degeneracy advocates as William Robertson. De Pauw's *Recherches philosophiques sur les Américains*—or at least sections of it—were finally translated and compiled in 1789 by an Englishman named Daniel Webb. Webb initially published just fifty copies of *A General History of the Americans . . . Selected from M. Pauw* for family and friends. A few years later, though, he produced a commercial copy of *A General History*, and it was through this edition that most non-Francophones, especially those in England, became familiar with its details.

In his translation of de Pauw, Webb used italics to set off his own additions from the words of de Pauw: "The additions are printed in italicks," he told his reader, "to distinguish them from the selections: that neither M. Pauw, nor the author of the additions, might be responsible for what was not his own."[19] This sort of comment might have led the reader to believe that Webb would be critical of de Pauw's ideas, but nothing could have been further from the truth. Webb was as staunch a supporter of the New World degeneracy idea as could be found anywhere, and he peppered his translation with comments in support of the author and his platform.

Uncritically accepting de Pauw's claim of cone-headed savages, Webb demurred, "Little indebted to Nature for his other endowments, the savage seems in this to retort her injustice, by defacing the fairest example of her art."[20] Though he thought de Pauw was too philosophical in places, Webb had the "greatest respect for the genius and learning" of his author. Webb himself may have found the New World cannibals discussed by de Pauw "disgusting,"[21] but noted that de Pauw "thought it, no doubt, his duty as an historian to undertake this task."[22] And, in case the reader was still uncertain about the power of de Pauw's ideas, the very last words of Webb's translation made this clear: "I cannot take leave of our author without possessing the highest esteem of his genius and erudition." And why? Because in de Pauw, Webb saw open-mindedness: he saw a man who criticized ideas "not open to proof" and a man who devised "ingenious conjectures in such as are not."[23] And it was through the eyes of Webb that de Pauw's invectives on the New World were introduced to the English reader.

In conjunction with Buffon's and Raynal's works, de Pauw's book on degeneracy, whether read in English, French, German, or Dutch, influenced some of the most well-known thinkers of the late eighteenth and early nineteenth centuries—including men in philosophy, literature, natural history, and biology.

In many European philosophical circles at the time, the idea of New World degeneracy was readily accepted. Voltaire, for example, thought well of de Pauw. Indeed, when the famed meeting took place between Benjamin Franklin and Voltaire, Franklin would no doubt have been dismayed to learn that the man he kissed in front of members of the French Academy considered Cornelius de Pauw, "a very capable man, full of wit and imagination: a little systematic, in truth, but someone capable of amusing and instructing one."[24]

The German philosopher Immanuel Kant (1724–1804), also thought well of the degeneracy hypothesis. Initially, Kant was fond of de Pauw's work not so much for its accuracy, but because he believed it to be thought-provoking. "In Pauw, even if nine-tenths of his material is unsupported or incorrect," Kant wrote, "the very effort of intelligence deserves praise and emulation, as making one think and not simply read thoughts."[25]

Through time, Kant began to believe that perhaps de Pauw was not wrong nine-tenths of the time—and that neither was Buffon, whom Kant admired very much. In 1775, Kant described Americans as a race "not yet properly formed, or half degenerate. . . . Their vital force is almost extinct, and they are too feeble for any sort of agricultural work."[26] His opinion remained unchanged after Jefferson's *Notes* was published. In his 1788 book, *On the Use of Teleological Principles in Philosophy*, Kant blamed the climate in America for producing a race "too weak for hard work, too indifferent to pursue anything, incapable of any culture."[27]

Kant's writings on the inferiority of the New World are primarily aimed at South American Indians. He, like many other advocates of the degeneracy theory, denied that Patagonian giants were evidence against it, for the simple reason that Patagonians were not giants—the New World, Kant believed, was simply not capable of producing such vigorous men. With respect to animal life, Kant argued that while the birds of America were very colorful, they sang poorly, and in an age of Prussian society where music was viewed as the height of culture, such animals were clearly degenerate.

Kant's ideas on degeneracy and birdsong were not original. The dearth of song among American birds was commented on by Buffon, who claimed that the damning combination of wet and cold weakened the birds' sexual organs, and hence they could not carry a tune. To add insult to injury, Buffon also noted that the poor birds of the New World really had little inspiration—just the harsh grunts of savages.[28]

The muted reputation of American birds can be traced back to Oliver Goldsmith's 1769 poem, *The Deserted Village*. The stage is the wilderness of Georgia, "that horrid shore . . . where poisonous fields . . . where the dark scorpion gathers death around," and "the rattling terrors of the vengeful snake," can be heard. What can't be heard, however, are the birds, for as Goldsmith somberly noted:

> Those blazing suns that dart a downward ray,
> And fiercely shed intolerable day;
> Those matted woods, where birds forget to sing.

The one exception was the mockingbird, who, as later described in Goldsmith's multivolume series, *A History of the Earth and Animated Nature,* sang so beautifully that "the deficiency of most other song-birds in that country is made up by this bird alone."[29]

Mockingbirds aside, Oliver Goldsmith and others likely assumed that American birds were mute because naturalists in Europe encountered such birds either as stuffed specimens or as lone individuals in small cages, where they were unlikely to sing.[30] In any case, this characterization of degenerate American birds was instilled in the European culture to the extent that Buffon, Kant, and many others who promoted the idea of New World degeneracy, both before and after Jefferson's *Notes,* made much of these birds that "forgot to sing."

Philosopher Georg Wilhelm Friedrich Hegel (1780–1831), who both studied and admired Kant, was also a vociferous advocate of the theory of degeneracy. Although he believed that the future might belong to America because of its youth, Hegel saw the current state of America as degenerate. Hegel was reluctant to admit that nature made "mistakes," but he saw no way around that conclusion when he looked at the New World. Nowhere was it more apparent to Hegel that "the impotence of nature brings about . . . impure presentation."[31]

Hegel, whose major works were written decades after Jefferson's *Notes,* saw nothing of beauty in the New World. The Old World was divided into three integrated sections, while the New World had a puny land mass connecting the north and south. The rivers in the New World flowed haphazardly compared to those of the Old World, and the mountain chains ran in the wrong direction. Following up on Buffon's claim that America was immature, Hegel went one step further, decrying that "America has always been and still shows itself physically and spiritually impotent."[32] Again echoing Buffon, when describing animal life in the New World compared to that in the Old, Hegel noted, "They are in every way smaller, weaker and more cowardly." And this inferiority applied to domesticated animals as well as wild ones, with Hegel the epicurean warning the reader that "a piece of European beef is a delicacy" compared to that found in America.[33] Like Kant, Hegel gladly paraded the mute birds of America as evidence of degeneracy. These birds would one day sing, but only when they lived in a land that no longer "resound[s] with almost inarticulate tones of degenerate men."[34]

In Kant and Hegel we see two of Europe's leading philosophers who adopted the degeneracy theory as their own, even after Jefferson published his reply to Buffon. The theory was making its way into the heart of European culture, *Notes on the State of Virginia* or not. Philosophers may have been some of the first to grasp onto the hypothesis, but they were not the last; it seeped into the popular literature of early nineteenth-century Europe as well.

References to New World inferiority made their way into the writings of major poets, such as John Keats (1795–1821). Keats's brother, George, and his sister-in-law had sailed for the United States in June 1818, and for the next sixteen months the poet despaired over receiving no word of their fates. As he anxiously awaited news of George and his wife, Keats tried to ease the tension by learning what he could of the America that was now their home. On April 28, 1819, Keats wrote of "reading lately . . . [William] Robertson's *America*," the very Robertson that Thomas Jefferson described as "a mere translator of the opinions of Monsieur de Buffon."[35] It was through Robertson's works that Keats came to know America.

When Keats read Robertson's *History of America* in 1819, it was his second time through these volumes, the first time being as a schoolboy.[36] Already despondent that George had left the civilized world of England for the wilds of America, Keats was convinced that his brother and sister-in-law had been swindled by the American naturalist Audubon in a deal that involved a wrecked boat, and that they might very well be dead (they weren't). In any case, Keats was predisposed to the idea of New World inferiority. And degeneracy, with America as "the hated land," permeates a poem called *Lines to Fanny* (also called *The Second Ode to Fannie Brawne*)[37]—a piece which Keats wrote the very same month he was reading Robertson:

Where shall I learn to get my peace again?
To banish thoughts of that most hateful land . . .
That monstrous region, whose dull rivers pour
Ever from their sordid urns unto the shore . . .

Iced in the great lakes, to afflict mankind;
Whose rank-grown forests, frosted, black, and blind . . .
There bad flowers have no scent, birds no sweet song,
And great unerring Nature once seems wrong.[38]

Literary analyst H. E. Briggs has made a compelling case that almost every line above comes directly from Keats's reading of Robertson's degeneracy diatribe on the New World.[39] The "hateful land" is a poetic rendition of Robertson's portrayal of mortified Europeans, who upon encountering the "desolate and horrid" New World, were "astonished. . . . It appeared to them waste, solitary and uninviting." The "monstrous region," and "pouring rivers" are tied to Robinson's descriptions of the "immense extent" of America, with rivers that "flow in such spacious channels, that, long before they feel the influence of the tide, they resemble arms of the sea." Keats's "iced in the great lakes, to afflict mankind" is an allusion to Robertson's "frigid zones," and his description of vast stretches of America under snow and ice. Again we see the songless birds of Buffon and Goldsmith, which Keats learned about in Robertson's description of American birds as "denied . . . [the] melody of sound."[40] Degeneracy is then summed up by Keats when he describes a land where "great unerring Nature once seems wrong."

Keats died of tuberculosis at age twenty-five, and while his work was well known in literary circles of the day, his real fame emerged later, during the Victorian era, when he was universally recognized as one of England's most brilliant poets. It remains an open question whether the Victorian readers of Keats's *The Second Ode to Fannie Brawne* would have recognized that the poet was writing about America—they may or may not have. But the larger issue is that degeneracy had now made its way into the works of one of England's most renowned Romantic poets.

The year before Keats died, Sydney Smith continued the European, degeneracy-based assault on American literature. In an 1820 article that Smith published in the *Edinburgh Review*, he savaged American literature in a way that made Keats looks tame. "During the thirty or forty years of their independence, they have done absolutely nothing for the Sciences, for the Arts, for Literature," Smith demurred. Then he posed a question that would have brought a smile to the faces of advocates of the theory of American degeneracy everywhere: "In the four quarters of the globe, who reads an American book? or goes to an American play? or looks at an American picture or statue?"[41]

Buffon's original ideas on New World inferiority, as well as Raynal's and de Pauw's embellishments, also colored the worldview of scientists for more than a century. As a case in point, Jean-Baptiste Lamarck (1744–1829), one of the first evolutionary biologists, was Buffon's protégé at the Royal Gardens.[42] While Lamarck's work itself did not focus on degeneracy, it emerged directly from the idea.

Lamarck, whose ideas predated Darwin's by decades, was the first biologist to posit a process that could lead to real evolutionary change, an idea that has been labeled "the inheritance of acquired characteristics." The basic idea was that traits that organisms develop during their lifetimes are somehow passed on to their offspring. For example, in his 1809 magnum opus, *Zoological Philosophy*, Lamarck argued that shorebirds that fed in muddy areas and stretched their muscles to raise themselves from the mud and search for food would produce offspring that had longer legs.[43] Today, we know that this is not the way that environment interacts with genetics, but Lamarck lived before Mendel's discovery of the laws of genetics, and, at the time, this was a reasonable—in fact, quite ingenious—hypothesis.

Lamarck's ideas on evolutionary change were spurred, in part, by Buffon's ideas on New World degeneracy. Buffon's theory provided a means by which the environment had a dramatic impact on organisms. The cold and wet of the New World had a debilitating effect on life there. What Lamarck did was add a mechanism through which these effects of climate would be passed down from generation to generation. If cold, wet environments stunted growth in one generation, the effects could now be felt in many subsequent generations by the inheritance of these acquired traits, adding yet more punch for the argument for degeneracy.

When a young Charles Darwin accompanied Captain FitzRoy on his now famous voyage (1831–1836) on the HMS *Beagle*, he was already familiar with Buffon's theory of degeneracy, and it had created a powerful expectation of what he would find in his travels. When the *Beagle* made port in the area of Patagonia, Darwin's journal entry noted that "the zoology of Patagonia is as limited as its flora"—precisely the sort of thing his reading of Buffon's degeneracy theory would have led him to expect. And yet, despite this, Darwin wrote that "it is impossible to reflect on the changed state of the American continent without the deepest astonishment." What caused such deep astonishment? His explorations in South America had led him to believe that "*Formerly* it must have swarmed with great monsters: now we find mere pigmies, compared with the antecedent, allied races." Darwin's thoughts immediately turned to the Count and his *Natural History*: "If Buffon had known of the gigantic sloth and armadillo-like animals, and of the lost Pachydermata, he might have said

7.2. HMS *Beagle*. Painting by Conrad Martens.

with a greater semblance of truth that the creative force in America had lost its power, rather than that it had never possessed great vigour."[44]

In Darwin's view, Buffon's assessment of the current state of this continent was not far from the mark, regardless of its origins. Darwin never accepted Buffon's reason for the state of the flora and fauna of South America—that a wet, cold environment inherently led to degeneracy in America. Instead, he argued that while the exact causes were difficult to assess, in South America, like everywhere else, change was due to the gradual effects of the process of natural selection. Despite this difference in explanation, Darwin never took issue with Buffon's reading of contemporary America as degenerate. Indeed, the reader of *The Voyage of the Beagle* is left with the distinct impression that Darwin came to Patagonia with the expectation of seeing a degenerate set of animals and plants (at least in terms of the number of species), and, though he encountered some fantastic geological sights there, he left in a similar state of mind.

Despite the long list of those who continued promulgating the degeneracy idea after Jefferson wrote *Notes*, his refutation of Buffon was not a complete failure, as a number of important and talented people continued the fight. Some of these individuals were personal friends of Jefferson; others were not.

One of the first Europeans to come to the aid of Jefferson in combating the degeneracy hypothesis was Filippo (Philip) Mazzei, whom Jefferson once described as a man of "solid worth; honest, able, zealous in sound principles. Moral & political, constant in friendship, and punctual in all his undertakings."[45] Mazzei, an Italian physician who came to the American colonies in 1773 to introduce viniculture techniques to Virginia, quickly befriended Jefferson (and over time Franklin, Adams, Washington, and many other founding fathers of the United States). Because of his staunch support for the cause of American independence, in 1779, Jefferson, Patrick Henry, and others asked Mazzei to return to Italy and serve as a foreign agent for Virginia.[46] Though captured by the British and imprisoned for three months on his way back home, Mazzei eventually made it back to begin his new mission.

Mazzei's tasks included obtaining loans for Virginia from the Grand Duke of Tuscany, as well as gathering political and military information for the colonies. He served as an agent from 1779 to 1783, but soon after found himself unemployed and in Paris. Fortunately for him, Jefferson had recently been appointed minister to France and had the perfect new job for Mazzei—to defend the American constitution against a recent attack on that document by the Abbé Mably. Mazzei vigorously embraced this charge, and what was ostensibly to be an essay evolved into a four-volume compendium entitled *Historical and Political Researches on the United States of America.*[47]

Years earlier, having seen all that could be cultivated in Virginia, Mazzei had noted that "America produces everything." And so, when he learned of the great uproar that Raynal's work was having across Europe, Mazzei used his books, which were published in 1788, to take issue with Abbé Raynal. Indeed, the entire third volume of Mazzei's work ended up being devoted to defending America against Raynal's "pompous" claims of degeneracy.[48]

Jefferson gave Mazzei access to all the information he had used to write *Notes,* and along with his own research on the matter, Mazzei intended to show the world that "there was little that was true from beginning to end" in Raynal's books.[49] Because Raynal had turned "mistakes into axioms and blinds his reader with his magical style," Mazzei saw it as his mission to "tear the veil away to show them how far they have strayed from the facts."[50] He dismissed Raynal's (and hence Buffon's) notion that climate could lead to smaller individuals and fewer species in the New World, and that domesticated animals degenerated when brought over to America. And Mazzei went directly after de Pauw, accusing him of being a confidence man who sprinkled enough enticements into his tale to keep his victims coming back for more. While "practically all of them [de

Pauw's ideas] rely on some authority," Mazzei warned his readers, "you have to read a number of statements in order to find a single one that is true."[51] De Pauw, Mazzei noted disdainfully, was the "only writer surpassing Abbé Raynal in the number of his errors about America."[52]

A year before Mazzei's book appeared in print, Jefferson was writing friends that many articles written on the United States were using Raynal as their source, and hence were "wrong exactly in the same proportion." He then encouraged these same colleagues to read Mazzei's books. "Good work, very good one . . . published here [France] soon," he wrote of these volumes. And Mazzei himself was described as "many years a resident of Virginia, is well informed, and possesses a masculine understanding."[53] Jefferson's fondness for Mazzei was also a function of their joint passion for epicurean delights and wine—shortly before Mazzei's books were published, he was still exchanging letters with his Italian friend on such matters as "the best grapes for drying . . . olives, capers, pistachio nuts and almonds."[54]

Mazzei himself was so confident that he had aided Jefferson's cause that he believed a colleague who told him that through his work, de Pauw "realized his own errors,"[55] and might make amends for promoting the idea of degeneracy. De Pauw made no such amends to Mazzei, or anyone else, and instead continued to reprint his ideas on New World inferiority.

In the world of literature, Lord Byron held a very different view of America and degeneracy than that of his friend Keats. Rather than America as "the hated land," Byron saw an Old World in decline. America was not doomed to a poor climate that led to degeneracy, but was instead a land of strong, free people. Byron portrayed Washington as "Yes—one— the first—the last—the best—The Cincinnatus of the West,"[56] and wrote that he "would rather have a nod from an American than a snuff box from an Emperor."[57] The poet admired South America as well, and even considered moving there, despite having read degeneracy literature that he saw as "violent and vulgar party production."[58]

Byron's thoughts on New World degeneracy are best depicted in his "Ode to Venice," where of Europe he writes,

Ye men, who pour your blood for kings as water,
What have they given your children in return?
A heritage of servitude and woes,
A blindfold bondage, where your hire is blows.

In contrast, America is presented to the reader as

. . . [o]ne great clime,
Whose vigorous offspring by dividing ocean
Are kept apart and nursed in the devotion
Of Freedom, which their fathers fought for, and
Bequeath'd—a heritage of heart and hand,
And proud distinction from each other land . . .
Full of the magic of exploded science . . .
Still one great clime, in full and free defiance . . .

Byron's depiction of the land of America as "one great clime" and its people as "vigorous offspring . . . full of the magic of exploded science . . . in full and free defiance" were not the only poetic attempts to attack the supporters of the degeneracy theory. Shelley, too, spoke of America as "a mighty people in its youth." America was

. . . like an Eagle, whose young gaze
Feeds on the noontide beam, whose golden plume . . .
Thy growth is swift as morn, when night must fade;
The multitudinous Earth shall sleep beneath thy shade.[59]

European defenders against the charge of American degeneracy also came from the world of natural history. In response both to Jefferson's *Notes,* which he quoted at length in his own writings, and to his own travels in America, Alexander von Humboldt took issue with the degeneracy ideas of Buffon, Raynal, and de Pauw.

Humboldt was familiar with Buffon's, Raynal's, and de Pauw's arguments for the degeneracy theory even before his five-year (1799–1904), six-thousand-mile natural history expedition to Central and South America. He had serious doubts about this idea, however, believing that it was "unphilosophical and contrary to generally acknowledged physical laws."[60] Claims of American degeneracy, Humboldt believed, had been generated to garner attention rather than to "comprehend, in one general view, the construction of the entire globe."[61] And once he arrived in the New World, he fell in love with it, writing to his brother, "I cannot repeat to you enough how very happy I feel in this part of the world."[62]

The beautiful animals and luxuriant tropical forests that he encountered further convinced Humboldt that the theory of degeneracy was a fraud. In particular, though he respected Buffon as a naturalist, he was angry that such a man would craft the myth of New World inferiority: "It would be superfluous for me to refute here the rash assertions of M. de Buffon. . . . These ideas were easily propagated, because . . . they flattered

7.3. Alexander von Humboldt had harsh words for Buffon, Raynal, de Pauw, and the theory of degeneracy. Painting by Friedrich Georg Weitsch.

the vanity of Europeans . . . When facts are carefully examined, naturalists perceive nothing but harmony where this eloquent writer announced discordancy."[63]

Humboldt had harsh words for Raynal and de Pauw, too. Raynal was inaccurate (he "disfigures the names of cities and provinces")[64] and his claims were dismissed as "totally destitute of truth."[65] Indeed, Humboldt was quick to point to another writer who "says that the descriptions in Raynal are in general no more to be relied on than any description in romance."[66] De Pauw was also inaccurate (and thus not to be trusted), and his depiction of "America as a land of swamps, hostile to the multiplication of animals," Humboldt wrote, is an instance of "absolute skepticism [taking the] the place of a healthy criticism."[67]

Humboldt had one last part in the argument over degeneracy. Recall that in *Notes*, Jefferson had used the mammoth as one of his pieces of evidence against Buffon's theory. Humboldt was well aware of this, and marshaled it in an attempt to meet President Jefferson in May 1804. On his return from South America, Humboldt had been invited to visit the United States. While there, he wrote a letter of introduction to the president, in hopes of setting up a personal meeting. In this letter, after expressing a "moral interest" in the United States and his "high admiration" for Jefferson, Humboldt wrote, " I would love to talk with you about a subject that you have treated so ingeniously in your work on Virginia." Jefferson's interest was no doubt piqued when he next read of Humboldt's own discovery of "mammoth teeth . . . in the Andes of the southern hemisphere."[68] The meeting was arranged, as was a tour of Charles Peale's Natural History Museum, which housed the North American mammoth remains,[69] and Jefferson and Humboldt became lifelong friends. Indeed, though he already owned at least one copy of *Notes on the State of Virginia*, Humboldt came to treasure an autographed copy Jefferson sent him years after their time together in Washington, D.C.[70]

The European battle over degeneracy was not settled by *Notes on the State of Virginia*. Jefferson's attempt to calm the swell of opinion coming from Europe was only moderately successful. There were those like Humboldt and Mazzei, and Shelley and Byron, who spoke up, but their voices had to contend with Kant and Hegel, Darwin and Lamarck, and Keats. The argument over New World inferiority also had unexpected consequences in the United States, where it helped shape an image of America and of Americans that continues to this day.

CHAPTER EIGHT

Extracting the
"Tapeworm of Europe"
from Our Brain

Thomas Jefferson did not intend to use *Notes on the State of Virginia*, or the giant moose, to shape the way Americans would see the land they inhabited or those with whom they shared that land, be they other people (Native Americans) or other species. But arguments often take on a life of their own, and that is what happened with the degeneracy argument.

After Jefferson made his case, and for the next seventy years, other Americans—from those publishing textbooks for schoolchildren, to poets and writers such as Washington Irving, Henry David Thoreau, and Ralph Waldo Emerson—would not only denounce the insidious idea of degeneracy, but would use it in ways that would have made Buffon, Raynal, and de Pauw cringe. For what happened was this: in countering the idea of New World inferiority, these writers and poets created a novel self-image for the United States and its inhabitants—America as a beautiful, vast, resource-rich region, and its inhabitants as healthy, hardworking people in tune with nature.

One way in which early American authors countered the degeneracy theory was through the use of school textbooks.[1] The most striking example of this tactic can be found in the textbooks of the Reverend Jedidiah Morse (1761–1826). A graduate of Yale and one of the most influential New England Congregationalist pastors during the Second Great Awakening, Morse was also a well-respected historian and geographer of his day, and a contributor to the first encyclopedia published in the United States.[2]

At the age of twenty-three, while a student of the Reverend Jonathan

Edwards, Jr., preacher-in-training Morse wrote a textbook for school-children.[3] His *Geography Made Easy* was the first geography book intro-duced into the classrooms of America, just a year after the final treaty ending the Revolutionary War was signed.[4] But it was Jedidiah Morse the historian who had the largest part in programming the minds of Ameri-ca's schoolchildren against the idea of New World degeneracy. The initial chapter of his popular 1790 textbook, *The History of America in Two Books*, is largely a refutation of the ideas of Buffon and de Pauw.[5] The first thing young students learned in history class was how utterly misguided such slander from Europe really was. Students read of the "extreme malignity of climate [that] has been inferred and asserted by de Pauw," and quickly learned that de Pauw had simply taken Buffon's ideas ("at whose fountain [he] has drank") and regurgitated them in even viler form.[6]

Morse spent a large part of chapter 1 of *The History of America in Two Books* quoting, at length, both Buffon and de Pauw, so students could see for themselves the sort of material that European philosophers were writ-ing about their country. Not only was Morse convinced that the idea of degeneracy was incorrect, but he told his young readers that even if cli-mate were responsible for the traits that Buffon and de Pauw believed existed, they had misapplied their own theory. Instead of degeneracy, from the "smaller size and less fierceness of [America's] animals," Buf-fon and de Pauw should have deduced "the gentleness and sweetness"[7] of America's climate.

Morse cataloged many instances in which supporters of the degen-eracy idea simply had their numbers wrong. Insects and vermin weren't more common in America—the Old World had just as many swarms of "disgustful insects" as the New.[8] And to the chagrin of the young adven-turers in the class, they learned that Louisiana was not, in fact, teeming with frogs weighing thirty-seven pounds. At the end of chapter 1, Morse explained to the impressionable minds reading his book that the degen-eracy theory was the product of the "ignorance or . . . studied forgetful-ness of . . . the Old continent."[9] Some fifty years later, James Fenimore Cooper would make reference to the preamble of a Colonial school char-ter that read: "Whereas the youth of this colony are found, by manifold experience, to be not inferior in their natural geniuses to the youth of any other country in the world."[10] Some lessons were important enough to be taught over and over.

It was not only in one-room schoolhouses that American children of the early republic would learn of Buffon and his followers' ideas on de-generacy. On the anniversary of independence each year, Fourth of July orators would make both direct and indirect references to Buffon and what they saw as his slanderous theory. With flags flowing in the breeze, young and old in the audience heard of Jefferson "stepping forward" to challenge Buffon,[11] and how their fathers and forefathers had been "sent from Heaven, from the storms and miseries of Europe, to dwell in this land of promise."[12] Even in the more austere headquarters of the Literary and Philosophical Society of New York, people—important people—were stepping up to publicly denounce the idea of degeneracy. In 1814, during a speech to this society, DeWitt Clinton—future governor of New York and candidate for president of the United States—began his discourse by railing against Buffon (by name) and his idea of degeneracy, imploring his audience to take note of the obvious. Yes, the United States might not yet match its European cousins in some of the arts or sciences, but

> there is nothing in the fixed operation of physical or moral causes, nothing in our origin, in our migration or in our settlement; nothing in our climate, our soil, our government, our religion, our manners, or our morals, which can attach debility to our minds or can prevent the cultivation of literature. . . . Surely sufficient reasons may be assigned without impeaching the character of our minds or degrading us in the scale of being.[13]

The response to the theory of degeneracy, and its eventual shaping of the American sense of self, is evident in many genres of early American writing. One such genre was the satirical poem, many of which were political in nature. The "Hartford Wits,"[14] a group of Connecticut intellectuals who had strong Federalist leanings, were well known for such poems. These biting satires would often appear in newspapers and, above and beyond their political overtones, were a much beloved form of entertainment in the years following the American Revolution.

One of the more famous of the Hartford wits was Jefferson's friend Joel Barlow, who upon reading *Notes on the State of Virginia* wrote to his colleague that he was overjoyed at "the idea of seeing ourselves vindicated from those despicable aspirations which have long been thrown upon us and echoed from one ignorant scribbler to another in all languages of Europe."[15]

Barlow, best known for his work *The Columbiad*,[16] used his biting wit to mock the degeneracy hypothesis in another of his poems called *The Anarchiad* (1787). In this poem, Buffon, Raynal, and de Pauw

> . . . scan new worlds with philosophic eyes . . .

What they saw through their "philosophic eyes," is a land where

. . . enfeebled powers of life decay
Where filling suns defraud the western day
Paint the dank, sterile globe, accurst by fate
Created, lost, or stolen from ocean late

For Barlow, these scans of America by the proponents of the degeneracy theory were done through a bizarre philosophical telescope with "inverted optics":

See vegetation, man, and bird, and beast,
Just by the distance squares in size decreased . . .
Huge mammoth dwindle to a mouse's size
Columbian turkeys turn European flies
Exotic birds, and foreign beasts, grow small
And man, the lordliest, shrink to least of all
While each vain whim their loaded skulls conceive.
Whole realms shall reverence, and all fools believe . . .

There, with sure ken, th' inverted optics show
All nature lessening to the sage De Pau[w]
E'en now his head the cleric tonsures grace,
And all the abbe blossoms in his face;
His peerless pen shall raise, with magic lore
The long-lost pigmies on th' Atlantic shore . . .

The Anarchiad was not the only satirical poem that attacked the degeneracy theory. David Humphreys, one of Barlow's coauthors on *The Anarchiad* and a former aide-de-camp to General Washington during the Revolutionary War, used his popular play *The Widow of Malabar* to the same effect:

And let philosophers say what they please
Your [sic] not grown less by coming o'er the seas . . .
Your victories won—your revolution ended
Your constitution newly made and mended . . .
Will make the age of heroes, wits and sages
The first in the story to the latest ages![17]

Through the textbooks of Morse, and the poems and plays of those like the Hartford Wits, the degeneracy hypothesis was being countered in the popular culture of early America. And while it was implicit in this

8.1. Joel Barlow, who used his sharp wit to mock the degeneracy theory in his satirical poetry. Painting by Robert Fulton.

debunking that America was equal—if not superior to—the Old World, the explicit sense in which accusations of degeneracy would lead Americans to see their country as a beautiful, vast, resource-rich region, and themselves as healthy, hardworking people in tune with nature had not yet developed. That would come in the work of Washington Irving. Often referred to today as the first genuinely American author, Washington Irving's prose would put to rest Raynal's claim that America had not produced one man of genius in the arts.

Irving was quite familiar with the writings of Buffon, mentioning the naturalist and his ideas in his *Salmagundi; Or, the Whim-whams and Opinions of Launcelot Langstaff, Esq.*, in *The Sketch-Book of Geoffrey Crayon*, and in his biography of naturalist Oliver Goldsmith. The first snipe at Buffon appeared in *Salmagundi* (1807–1808), which was published after Irving had spent 1804–1806 traveling around Europe. Here, Irving noted that Buffon (among others) did not know that the eagle was a singing bird.[18] Though a passing reference—it took up all of two lines in *Salmagundi*—this first instance of Irving calling Buffon to task is significant in that it deals with one of the most irritating claims made by degeneracy theorists, namely that American birds didn't sing, and that this was evidence of the inferiority of life in the New World. Irving was something of an expert on the question of singing birds and degeneracy; indeed, he went on to write a book-length biography of Oliver Goldsmith—the poet/naturalist who had written *The Deserted Village* poem immortalizing American birds "that forgot to sing."[19]

In *The Sketch-Book of Geoffrey Crayon* (1819–1820), which Irving wrote while living in England, degeneracy became a more prominent theme, through Irving's biting satirical wit. The American Geoffrey Crayon, who in some ways was a fictionalized version of Irving himself, possessed an "earnest desire to see the great men of the earth." And for that, Crayon turned to the Old World. Yes, there were "great men in America," but not great enough. He was anxious "to see the great men of Europe." The reason was clear, as Crayon claimed that he had "read in the works of various philosophers, that all animals degenerated in America, and man among the number. . . . A great man of Europe . . . must therefore be as superior to a great man of America, as a peak of the Alps to a highland of the Hudson." Only one thing remained: "I will visit this land of wonders, thought I," Crayon wrote, "and see the gigantic race from which I am degenerated."[20]

Though his character Crayon was being facetious, Irving did have great respect for the minds of Europe, and believed that "Europe held forth all the charms of storied and poetical association. . . . My native country was full of youthful promise; Europe was rich in the accumulated treasures of age. Her very ruins told the history of the times gone by, and every mouldering stone was a chronicle." Visits to the Old World were a useful way to forget the "commonplace realities of the present, and lose myself among the shadowy grandeurs of the past."[21] At the same time that he admired some aspects of European culture, Irving had an unabiding love for the beauty of the American continent.

The distinction between the "storied charm" of the castles of the Old World and the grandeur and vastness of the American continent was a repeating theme in many of Irving's works. He used Geoffrey Crayon's *Sketch-Book* to highlight the beauty of America in no uncertain terms: "I visited various parts of my own country; and had I been merely a lover of fine scenery, I should have felt little desire to seek elsewhere its gratification, for on no country had the charms of nature been more prodigally lavished."[22] These charms centered on the land itself: "Her mighty lakes,"

8.2. Washington Irving mentions Buffon and the theory of degeneracy in such works as *The Sketch-Book of Geoffrey Crayon*. Sketch from *Washington Irving* by Charles Dudley Warner (1881).

Irving (through Geoffrey Crayon) noted, "her oceans of liquid silver; her mountains, with their bright aerial tints; her valleys, teeming with wild fertility; her tremendous cataracts, thundering in their solitudes; her boundless plains . . . her broad, deep rivers, rolling in solemn silence to the ocean; her trackless forests, where vegetation puts forth all its magnificence; her skies, kindling with the magic of summer clouds and glorious sunshine." America was second to none in terms of its natural history. "No, never need an American look beyond his own country," the reader was instructed, "for the sublime and beautiful of natural scenery."[23]

Irving did more than use *The Sketch-Book of Geoffrey Crayon* to counter degeneracy claims and then paint a picture of America as a natural history wonder. In a chapter entitled *Traits of Indian Character*, he directly attacked Buffon's, Raynal's, and de Pauw's claims that American Indians were degenerate.[24] Irving began by briefly quoting from the story of the Indian Chief Logan, whom Jefferson had used so powerfully in *Notes on the State of Virginia.* Throughout the course of the chapter, Irving recounted other stories depicting Indian valor and dignity, bemoaning, "Can any one read this plain unvarnished tale without admiring the stern resolution, the unbending pride, the loftiness of spirit that seemed to nerve the hearts of these self-taught heroes . . . ?"[25]

The Native Americans depicted in *The Sketch-Book* are anything but degenerate. "There is something in the character and habits of the North American savage," Irving wrote, "taken in connection with the scenery over which he is accustomed to range, its vast lakes, boundless forests, majestic rivers, and trackless plains, that is, to my mind, wonderfully striking and sublime."[26]

"In discussing the savage character," Irving noted, "writers have been too prone to indulge in vulgar prejudice and passionate exaggeration, instead of the candid temper of true philosophy."[27] The mischaracterization of Native Americans as degenerate by so many was due to ignorance: "If we would but take the trouble to penetrate through that proud stoicism and habitual taciturnity which lock up his character from casual observation, we should find him linked to his fellow-man of civilized life by more of those sympathies and affections than are usually ascribed to him."[28]

Before the white man came, the Indian was even more magnificent. "How different was their state while yet the undisputed lords of the soil!" wrote Irving. "Their wants were few and the means of gratification within their reach. . . . Such were the Indians whilst in the pride and energy of their primitive natures."[29] Once the Europeans arrived, all that changed, and the Indians, Irving argued, responded as best they could: "If courage intrinsically consists in the defiance of danger and pain, the life of the Indian is a continual exhibition of it."[30] That was probably not going to be enough, though. Irving saw Indian culture as doomed: "The eastern tribes have long since disappeared . . . and such must, sooner or later, be the fate of those other tribes which skirt the frontiers. . . . In a little while, and they will go the way that their brethren have gone before. . . . They will vanish like a vapor from the face of the earth; their very history will be lost in forgetfulness . . . or if, perchance, some dubious memorial of them should survive, it may be in the romantic dreams of the poet, to people in imagination his glades and groves, like the fauns and satyrs and sylvan deities of antiquity."[31]

The Sketch-Book of Geoffrey Crayon, which included such stories as "Rip Van Winkle," became a huge success in the United States, and through it many readers who may not have been familiar with the ideas of degeneracy were now informed on the subject. Irving, the first great man of letters in America, used the arguments around degeneracy to establish the notion of America as a land of vigor and beauty, and Americans as noble—a land, and a people, antithetical to that described by Buffon, de Pauw, and Raynal.

A hundred years after Buffon began publishing *Natural History,* Thoreau still felt it necessary to demonstrate how utterly wrong the Count's ideas on the degeneracy of the New World were. Thoreau was a passionate natural historian, keeping detailed journals of what he saw around him. Eventually his passion for natural history evolved into an interest in conservation and preservation of natural habitats, reflected in such essays as "Natural History of Massachusetts," "Walden," "A Winter Walk," "Autumnal Tints," "The Maine Woods," "Wild Apples," "Walking," and many others.[32] It was in "Walking" that Thoreau presented his thoughts on Buffon and degeneracy.

Thoreau's opening sentence in "Walking"—"I wish to speak a word for Nature"[33]—set the tone for what was to follow. This essay is an ode to sauntering through the woods and absorbing the beauty of nature—in Thoreau's case, the nature of New England—where he could "walk ten, fifteen, twenty, any number of miles . . . without going by any house, without crossing a road except where the fox and the mink do."[34] Such walks, and the enlightenment they produce, come "only by the grace of God . . . a direct dispensation from Heaven."[35]

On his walks, which he described as akin to the "migratory instincts in birds and quadrupeds," Thoreau was always drawn westward and "not towards Europe" in his attempt "to forget the Old World."[36] He wanted not just to forget the Old World, but to praise the natural history wonders of the New World: "Where on the globe," Thoreau asks the reader of "Walking," "can there be found an area of equal extent with that occupied by the bulk of our States, so fertile and so rich and varied in its productions, and at the same time so habitable . . . as this is?"

In "Walking," Thoreau quotes Sir Francis Head, governor of Canada, who seems to have captured Henry David's own sense of the magnificence of the New World:

> In both the northern and southern hemispheres of the new world, Nature has not only outlined her works on a larger scale, but has painted the whole picture with brighter and more costly colors than she used in delineating and in beautifying the old world. . . . The heavens of America appear infinitely higher, the sky is bluer, the air is fresher, the cold is intenser, the moon looks larger, the stars are brighter, the thunder is louder, the lightning is vivider, the wind is stronger, the rain is heavier, the mountains are higher, the rivers larger, the forests bigger, the plains broader.[37]

8.3. Henry David Thoreau discusses Buffon and degeneracy in his essay, "Walking." This is the Dunshee ambrotype of Thoreau.

This passage is followed by a clear verdict from Thoreau: "This statement will do at least to set against Buffon's account of this part of the world and its productions."[38] The New World was healthy and vigorous, and this led to a sort of serenity, for "in the East-Indian city of Singapore, some of the inhabitants are annually carried off by tigers; but the traveller can lie down in the woods at night almost anywhere in North America without fear of wild beasts."[39] Even a swamp in America—that vile place that Buffon had thought so critical in explaining degeneracy in the New World—was a thing of beauty for Thoreau, who described it as "a sacred place—a *sanctum sanctorum*."[40]

Buffon, Raynal, and de Pauw, it seems, had gotten it all backward. "Climate does . . . react on man," Thoreau readily admitted, and that is what made America and Americans great. "There is something in the mountain-air," Thoreau noted, "that feeds the spirit and inspires. Will not man grow to greater perfection intellectually as well as physically under these influences?"[41] Surely the citizens of the United States would, he told his reader, "be more imaginative, . . . our thoughts will be clearer, fresher, and more ethereal, as our sky, our understanding more comprehensive and broader, like our plains, our intellect generally on a grander scale, like our thunder and lightning, our rivers and mountains and forests, and our hearts shall even correspond in breadth and depth and grandeur to our inland seas."[42] This was a deep, ingrained feeling, and Thoreau was certain that he was right: "Else to what end does the world go on," he asked, "and why was America discovered?"[43]

Like Irving, Thoreau also took issue with degeneracy's implications for Native Americans (as did many of his fellow Transcendentalist colleagues). Though he felt that Indian civilization featured some "fixed habits of stagnation,"[44] in general, Thoreau adopted a view of Indians similar (but not identical) to Rousseau's "noble savage," writing in his journal that "the charm of the Indian to me is that he stands free and unconstrained in Nature, is her inhabitant and not her guest, and wears her easily and gracefully."[45] Indeed, over the last decade of his life, Thoreau filled his journals with almost three thousand pages of notes on Native American anthropology for a book he may have been planning (but which he never wrote).[46]

Ralph Waldo Emerson also discussed degeneracy, although not as directly as his friend and colleague Thoreau. Unlike Thoreau, Emerson never mentioned Buffon by name, although references to the effect of climate and degeneracy appear throughout a number of his works. In his essay "Civilization," he spelled this out clearly: "The highest civility has never loved the hot zones. Wherever snow falls, there is usually civil freedom. Where the banana grows, the animal system is indolent and pampered at the cost of higher qualities: the man is sensual and cruel."[47] On the surface, this power of climate to shape life has Buffonian overtones, except now it is in the cold zones, where snow falls, not warm and dry climes, that civility thrives. A more important difference, though, is that Emerson does not see the effects of climate as inevitable. "This scale is not invariable. High degrees of moral sentiment control the unfavorable influences of climate."[48]

Emerson respected the contributions of the great European thinkers of the past,[49] but was always looking to the here and now and to the future: "Genius looks forward," he wrote; "the eyes of man are set in his forehead, not in his hindhead."[50] With respect to literature, Emerson saw the United States as the future—as on the cusp of disproving Abbé Raynal's claim regarding the lack of genius in the New World. Yes, it was true that "we have yet had no genius in America," Emerson noted, but the past is the past, and what's more, a search for individual greatness missed the point: "America is a poem in our eyes; its ample geography dazzles the imagination, and it will not wait long for metres."[51] The land and the people themselves represented genius.

8.4. Ralph Waldo Emerson asked his readers, "Can we never extract this tapeworm of Europe from the brain of our countrymen?" Portrait by David Scott.

Emerson "read with joy some of the auspicious signs of the coming days, as they glimmer already through poetry and art."[52] It was time to move away from the cultures of the past: "Can we never extract this tapeworm of Europe from the brain of our countrymen?"[53] Emerson implored his readers. Surely they could, and the time to do so was now: "Our day of dependence, our long apprenticeship to the learning of other lands, draws to a close," Emerson told the Phi Beta Kappa Society in 1837. "The millions that around us are rushing into life cannot always be fed on the sere remains of foreign harvests."[54]

After the days of Thoreau and Emerson, Buffon's theory of New World degeneracy faded into obscurity. Likewise, its protagonists—men like Raynal, de Pauw, and Robertson—are mentioned only among a very small set of scholars who continue to be fascinated by them.

Modern lovers of art can readily purchase exquisite prints from Buffon's *Natural History* everywhere from galleries on Park Avenue to auctions on eBay, and Buffon's encyclopedic work is still considered to be the starting place for the modern study of natural history. As for the Count himself, he is still regarded as one of the torchbearers of the Enlightenment. But his idea that the New World, and in particular, America, was inherently inferior, has all but vanished.

If Jefferson had survived to the middle of the nineteenth century as a centenarian, he would have been pleased. It is true that the giant moose didn't have the immediate effect he had hoped for, what with Buffon's unexpected death and his unfulfilled promise to make corrections to *Natural History*. But Jefferson's *Notes on the State of Virginia*, and the physical evidence—moose, mastodon, panther skin, and so on—which he presented to Buffon—were the key weapons in the arsenal of ideas used in the hundred-year fight against the degeneracy hypothesis. And while it was not Jefferson's goal to use his arguments to create an American ethos that centered on a healthy land and a vigorous people, he would surely have been pleased at that unexpected byproduct.

Why the degeneracy theory vanished when it did is difficult to say—certainly, more pernicious, equally unfounded speculations have lasted longer than the one hundred years between Buffon's original ideas and the mid-nineteenth century. Part of the reason may have been a natural decay. By the 1850s, the key players were all long dead, and the loss of such dramatic personalities would naturally result in a loss of interest in what they were fighting about.

While Jefferson and his compatriots knew the degeneracy theory to be false from its inception, by the mid-nineteenth century it would have been patently obvious to almost all the players on the world stage that America was not inherently inferior. People moved more freely between Europe and America and could see for themselves that what Buffon, Raynal, de Pauw, and others had said was false. The facts on the ground were clear. Natural history had advanced significantly since the time of Buffon, and there was no evidence of American degeneracy. "Science" itself was being recognized as a profession in the mid-1800s, and there was no scientific reason to expect that climatic differences should lead to one part of the world being superior to another.

The economic facts too spoke against degeneracy. Though it was soon to be engulfed by a bloody civil war, America was clearly a success, having grown by leaps and bounds, and having used its natural resources to create the beginnings of an economic juggernaut.[55] In the face of all that, support of the theory of degeneracy—in the sense that Buffon and his followers used the term—diminished and then disappeared.

ACKNOWLEDGMENTS

Christie Henry, my editor at the University of Chicago Press, has championed this project from the start. Christie's editorial advice and friendship have been priceless, and her ability to convey the feeling that "all is well" to a type-A author has made all the difference.

I'm indebted to five anonymous reviewers, who provided insight, criticisms, and suggestions. In addition, the following individuals also read all, or parts, of the manuscript and provided helpful comments along the way: Matt Druen, Jerram Brown, Ryan Earley, Marc Bekoff, and Henry Bloom. As always my wife and helpmate, Dana, has read the entire manuscript more times than I can count, and her thoughts and suggestions have made the book that much better. Lastly, thanks to my son Aaron, for providing some relaxation during the book-writing process by getting me to Slugger Field all those nights to watch the Louisville Bats win their AAA division.

NOTES

Preface

1 Thomas Jefferson to John Sullivan, January 7, 1786 (Library of Congress Jefferson collection).
2 Briggs calls Keats's poem "the second ode to Fannie Brawne" (Selincourt 1905; Briggs 1944). This poem is also sometimes referred to as "What Can I Do to Drive Away."
3 Thoreau 1913, 173.

Notes to Chapter 1

1 Masterson 1946.
2 Brickell 1737.
3 du Pratz 1758, 2:264.
4 Dumont 1753.
5 Thomas Young in the *Essex Gazette*, July 10–17, 1770, 203.
6 Unlike the other travelers' tales mentioned earlier, this tale was from a fiction book (Chetwood 1720).
7 Dupree 1957.
8 Dudley 1720–1721.
9 Also see Barton 1803.
10 Hindle 1956; Greene 1984.
11 Dupree 1957, 7; Greene 1984.
12 Parrish 2006, 131.
13 Alexander Garden to John Ellis, November 19, 1764, Linnaean Society Correspondence, London.
14 Muhlenberg had a herbarium of 5,000 specimens, and in 1815 he would publish a catalogue of 3,670 species. Cutler authored *An Account of Some of the Vegetable Productions Naturally Growing in This Part of America, Botanically Arranged* (Humphrey 1898). Henry Muhlenberg to Manasseh Cutler, November 12, 1792.
15 Manasseh Cutler to Gustav Paykull, February 14, 1799.
16 Porter 1986.

17 Darwin 1859.
18 Ruse 2003.
19 Plato used this term in *Timaeus*.
20 Aristotle too saw living forms as designed, but did not turn to a discrete supernatural being as designer. For Aristotle, design was inherent in nature.
21 Case-Winters 2000.
22 Galen made this argument in his *On the Usefulness of the Parts of the Body*.
23 Especially evident in Hume's *Dialogues Concerning Natural Religion*.
24 In Bacon's *Advancement of Learning* (1605), 1:16.
25 Men such as Robert Boyle, William Derham, Henry More, and Ralph Cudworth (Gillespie 1987).
26 Ray 1717, preface (n.p.).
27 Ibid., 29.
28 Holmes 1940.
29 Mather 1693, front matter (n.p.).
30 Cotton Mather's *The Christian Philosopher* included some of the writings in his *Curiosa Americana*, a series of natural history essays he sent to the Royal Society between 1714 and 1723 (Beall 1961; Hornberger 1935).
31 Jeske 1986.
32 Mather [1721] 1815, 7.
33 Ibid., 5.
34 Ibid.
35 Mather [1721] 1815, 193.
36 Ibid., 195.
37 Ibid., 202.
38 Ibid., 198.
39 Ibid., 8.
40 Ibid., 194.
41 Natural theology is sometimes referred to as physico-theology (Paley 1802).
42 Nature as evidence of design was the dominant view among European naturalists of the era as well (Roger 1997).
43 Locke did not consider himself a Deist, but his writings are widely regarded as critical to the philosophy of Deism (Martin 1952).
44 Thomas Jefferson to John Adams, April 11, 1823 (Library of Congress, Jefferson collection).

Notes to Chapter 2

1 Gerbi 1973, 4.
2 There is some debate on the exact number of volumes in this work. Buffon was responsible for the first thirty-six volumes (which included seven "Supplements"). An additional eight volumes on fish and cetaceans were the work of Lacépède, and they are sometimes counted as part of *Natural History: General and Particular*.
3 Buffon 1749–1804, 2:28; Roger 1997, 133.
4 Piveteau 1952, 125–32; Roger 1997, 426–32.
5 Buffon was admonished by the Faculty of Theology of Paris for his claims (in volumes 1–3) that the waters of the seas produced mountains and valleys, the earth was created when a comet hit the sun, and the sun would eventually burn. The faculty was also upset by Buffon's definition of "truth" and his discussion of the soul. In a letter dated January 15, 1751, The faculty wrote:

> Sir: . . . When you learned that the *Natural History* of which you are the author was one of the works which had been chosen by order of the Faculty

of Theology to be examined and censured as containing principles which
are not in conformity with those of religion, you declared to him that you
had never had the intention of deviating in such matters, and were disposed
to satisfy The Faculty on each article which it might find reprehensible in
the work under question. Monsieur, we are not able to give adequate praise
to such a Christian resolution; and in order to afford you the opportunity to
execute your resolves, we forward to you the propositions extracted from
your work which have appeared to us to be contrary to the belief of the
Church. With complete esteem, we have the honor of being, Sir, Your most
humble and obedient servants, The Deputies and Syndic of the Faculty of
Theology of Paris.

Following this was a list of fourteen objections. In his March 12, 1751, reply, Buffon
wrote: "I declare: First: That I have never had any intention of contradicting the text
of Scriptures: that I believe quite firmly all that is related there concerning creation. . . .
I abandon that which, in my book, concerns the formation of the earth, and, in general,
all that which may be contrary to the narration of Moses, having only presented my
hypothesis on the formation of planets as a pure philosophical conjecture." Buffon then
offered to publish his entire response to the Faculty of Theology's queries in the next
volume (4) of *Natural History*. For English versions of the exchange between Buffon and
the Syndic see Lyon and Sloan 1981.

6 Buffon acknowledged God as he who "gave impetus" to the universe.
7 Buffon went so far as to congratulate himself for demolishing the idea of a nonhuman
soul, and at the same time for "having proven the spirituality of his [man's] soul. (Buffon 1749–1804, 2:444; Roger 1997, 162).
8 Farber 2000.
9 The French rank of Count translates roughly to the English title of earl.
10 For a masterful biography of Buffon, see Roger 1997. Also see: Fellows and Milliken
1972; Lyon and Sloan 1981.
11 Roger 1997, 5.
12 Including Gabriel Cramer, a prodigy not much older than Buffon himself, who already
held the post of professor of mathematics at the Geneva Academy.
13 In later years, Buffon would modify this game into one in which players bet on a needle
landing on the crack of a floor, and he would become famous for his solution (called
"Buffon's Needle").
14 At the time, the academy was divided into six classes: geometry, astronomy, mechanics,
anatomy, chemistry, and botany.
15 Buffon to LeBlanc, February 1738, from Buffon 1860, 1:33–34; Roger 1997, 30.
16 At an annual salary of 1,200 livres.
17 From the preface to Buffon's translation of Hales (Roger 1997, 26).
18 Also known as the intendant of the Royal Gardens.
19 Jean Hellot.
20 From Buffon 1860, 1:41–42; Roger 1997, 45.
21 3,000 livres per year.
22 There were, however, a few breaks along the way that allowed Buffon to pursue
other interests. In one, he would get into something of a competition with Benjamin
Franklin over who discovered electricity first, and in a second, he invented "burning
mirrors" that could set all sorts of objects on fire. This last interest was inspired by a
story that Archimedes set fire to Roman vessels by using a series of concave mirrors.
Rene Descartes argued that these mirrors would have to be extremely large to work,
but Buffon built much smaller versions (the largest measured six feet on one side) and
was able to burn wooden buildings at distances between ten and two hundred feet and
melt iron at ten feet. Buffon gave several demonstrations of these burning mirrors,

including one where he presented the king with one of his mirrors. In admiration, The *Mercure* newspaper published a poem in his honor:

> Buffon! There is nothing that does not cede
> To your ingenious efforts.
> What! The miracles of Archimedes
> Are merely the games of a studious leisure for you.

But these were sideshows for Buffon: *Natural History: General and Particular* was all he truly cared about from virtually the moment he took over the Royal Gardens.

23 Buffon 1749–1804, 1:30; Roger 1997, 89.
24 Mornet 1910; Loveland 2001; Loveland 2004.
25 *Journal de Trévoux*, September 1749, translated in Lyon and Sloan 1981.
26 This typically entailed reprinting Jefferson's replies to Buffon. *New Jersey Journal*, March 11, 1787; *Massachusetts Gazette*, May 8, 1787; *Massachusetts Centinel*, May 8, 1787; *Pennsylvania Packet*, June 13, 1787; *Maryland Herald*, September 4, 1800; *State Gazette of South Carolina*, July 29, 1793.
27 *State Gazette of South Carolina*, July 29, 1793.
28 Buffon 1860, 2:615; Roger 1997, 433. Neither Buffon's status as an intellectual celebrity nor the influence of *Natural History* among the masses ended with his death. In the introduction to his *Back to Methuselah*, George Bernard Shaw quipped that "every literate child . . . knew Buffon's *Natural History* as well as Esop's [*sic*] Fables" (Farber 1998).
29 Buffon 1749–1804, 1:30; Roger 1997, 89.
30 Buffon 1749–1804, 9:127; Roger 1997, 308.
31 Buffon 1749–1804, 1:4; Roger 1997, 83.
32 Loveland 2006.
33 Roger 1997, 142.
34 Like that of the famed naturalist Réaumur.
35 Buffon 1749–1804, 1:31–32; Roger 1997, 87.
36 Loveland notes that in addition to Buffon's writing ability, two other factors made *Natural History* a literary success. First, that natural history was particularly suited for commercialization (or "vulgarization," depending on one's perspective), as "many of its usual subjects—animals, flowers, precious stones, and so on—came with symbolic and literary prehistories which helped their packaging for the polite audiences." Buffon was a master at tapping into this prehistory. Second, Buffon was quite adept at using titillating language in his descriptions of the natural history of mating systems, luring in the reader without appearing crude. So titillating was some of this language that Marie-Jeanne Philipon forced herself to skip parts of *Natural History* to "preserve her moral dignity." Most readers, though, were not as upstanding, and savored, rather than skipped, these sections (Roland 1976, 46; Loveland 2001).
37 Buffon 1749–1804, 7:39–42; Barr 1792, 6:145–49.
38 Cuvier 1800; Blanckaert 1993.
39 Grimm 1877–1882, 2, 163–71.
40 Rousseau to Du Peyrou, November 4, 1764.
41 Even Buffon's modern biographer, Jacques Roger, who generally viewed his subject in a positive light, would recognize that Buffon "wrote too well and seduced the general public" (Roger 1997, xiii).
42 Jefferson [1787] 1999, 68. The 1999 reprint is based on a 1787 edition published by John Stockdale.
43 Chinard 1947; Glacken 1967; Gerbi 1973; Commager 1977; Roger 1997; Semonin 2000; Roger 2005.
44 Gerbi 1973.
45 Aristotle discusses this in *Politics*, Volume 7.
46 In a debate over spontaneous generation (Gerbi 1973, 8).

47 Gerbi 1973, xv.
48 Ibid., 40.
49 Echeverría 1957.
50 Buffon 1749–1804, 9:56–59; Barr 1792, 7:5–7.
51 Buffon 1749–1804, 9:72; Barr 1792, 7:15.
52 Buffon 1749–1804, 9:62; Barr 1792, 7:9.
53 Buffon 1749–1804, 9:68; Barr 1792, 7:13.
54 Buffon 1749–1804, 9:75–76; Barr 1792, 7:17–18.
55 Buffon 1749–1804, 9:70–71; Barr 1792, 7:14.
56 Buffon 1749–1804, 9:86; Barr 1792, 7:27.
57 Buffon 1749–1804, 9:86; Barr 1792, 7:27.
58 Buffon 1749–1804, 9:97; Barr 1792, 7:34.
59 Buffon 1749–1804, 9:100; Smellie's translation of Buffon 1781, 5:125.
60 Buffon 1749–1804, 9:101–2; Barr 1792, 7:38.
61 Buffon 1749–1804, 9:105–6; Barr 1792, 7:41.
62 Buffon 1749–1804, 9:103–4; Barr 1792, 7:39.
63 In addition to humidity, temperature, and the actions of the Native American popula-
    tion, Buffon believed that other factors could also affect degeneration. These included
    poor food and the lack of a sufficient number of males available for mating.
64 Buffon 1749–1804, 9:106–7; Barr 1792, 7:43.
65 Buffon 1749–1804, 9:109; Smellie 1781, 5:135.
66 Buffon, *Essay on Moral Arithmetic*, 1777, as translated in Lyon and Sloan 1981, 59.
67 Ibid., 61.
68 We will ignore the small probability that animals in both places are exactly the same
    size.
69 In 1959, at the hundredth anniversary celebration of the Société d'anthropologie de
    Paris, a medal was cast in Buffon's likeness. For more on Buffon and anthropology see
    Topinard 1883; Cunningham 1908; Blanckaert 1993.
70 This is known as the monogenesis hypothesis.
71 Roger 1997.
72 Buffon 1749–1804, 14:31–32; Roger 1997, 259.
73 Buffon 1749–1804, 3:435; Roger 1997, 175.
74 Blanckaert 1993, 33.
75 Buffon 1749–1804, 3:528–29; Roger 1997, 178.
76 Buffon 1749–1804, 14:314–15; Smellie 1781, 7:394.
77 Pagden 1982.
78 Buffon 1749–1804, 9:104–6; Smellie 1781, 5:128–30.
79 Buffon 1749–1804, 9:110; Smellie 1781, 5:135.
80 Buffon 1749–1804, 15:455–56; Gerbi 1973, 14.
81 Conklin 1947.
82 Chinard 1947, 41.
83 Roger 2005.

Notes to Chapter 3

1 Translation from de Pauw 1806, 17–18.
2 Biographic information on de Pauw can be found in Church 1936; Beyerhaus 1926.
3 In 1757, when he was eighteen, Cornelius may have fathered a boy named Charles.
   De Pauw never married, and his paternity here is the subject of some debate. It is
   worth mentioning, however, that Charles was supposedly sent to Paris to be educated,
   whereupon he met Lafayette, became enamored with the American Revolution, and
   eventually fought in the War of Independence (Church 1936, 179).
4 In a city called Luttich, also known as Liège.

5 In later years, de Pauw would publish other books on the history of the Egyptians, Chinese, and Germans.
6 This book went into its eleventh printing in 1799 (Roger 2005, 24).
7 De Pauw 1806, 13.
8 Ibid., 21.
9 Ibid., 27–28.
10 Ibid., 34–35.
11 Ibid., 21.
12 Ibid., 21–22.
13 Ibid., 34–35.
14 Ibid., 2.
15 Ibid., 23.
16 Ibid., 2.
17 Ibid., 17.
18 De Pauw 1768b, 2:108; Roger 2005, 19.
19 De Pauw 1806, 160; Roger 2005, 20.
20 De Pauw 1806, 76.
21 Ibid., 48.
22 Ibid., 63.
23 Ibid., 114.
24 Gerbi 1973.
25 De Pauw 1806, 17.
26 Ibid., 18.
27 Ibid.
28 Ibid.
29 Ibid., 27.
30 Ibid., 82.
31 Ibid., 114.
32 De Pauw, in his 1776 *Supplément de l' encyclopédie* article on America. See Chinard 1947; Commager 1977; Gerbi 1973.
33 Chinard 1947, 35.
34 Echeverría 1957; LaCorne 2005.
35 Raynal 1770; 1784.
36 Salmon 1976; Moore 2005.
37 Raynal served as editor from 1750 to 1754 (Salmon 1976).
38 Ibid., 110.
39 Tolchard 1957; Womack 1970; Salmon 1976.
40 As well as Edward Gibbon (author of *The Decline and Fall of the Roman Empire*) and Horace Walpole (Moore 2005, 20.)
41 The Dutch publication date of 1770 and the French date of 1772 have caused much confusion, with some researchers listing 1770 as the publication date for the first edition, and others 1772. I shall use 1770 as the date of the first edition.
42 The exact number depends on whether one counts pirated versions, and how one defines a new "edition" (Moore 2005, 17).
43 Sorel 1897, 55; Wolpe 1957, 8; Womack 1970, 4; Salmon 1976, 109.
44 Lusebrink and Strugnell 1995, 13.
45 Collaborators included d'Holbach, Grimm, Naigen, Galiani, Roux, Pechmeja, Deleyre, and Darcet (Duchet 1991; Salmon 1976, 113; Moore 2005, 21–22).
46 The *Encyclopédie, ou Dictionnaire raisonné des sciences, des arts et des métiers, par une société de gens de lettres.*
47 Later editions of *Philosophical and Political History* also extensively cited Thomas Paine's *Common Sense* (Salmon 1976; Duchet 1991; Moore 2005).

48 Raynal 1784, 2:1. Translated by J. O. Justamond. I shall refer to this as Raynal 1784.

49 Tolchard 1957, 95; Danzer 1974; Lusebrink and Mussard 1994; Niklaus 1995; Moore 2005, 73. Lusebrink and Mussard argue that the judges would have crowned Chastellux's prodiscovery essay, *Discours sur les avantages ou les désavantages pour l'Europe de la découverte de l'Amérique*, as the winner, but he refused to enter the essay. Moore suggests that Chastellux did not enter his essay because he was a member of Académie française, and tradition had it that members did not enter such contests.

50 Danzer 1974; Wood 1982. Danzer speculates that a ninth essay was written, but not submitted, by American Jeremy Belknap. This essay was published in May 1784 in *Boston Magazine*.

51 Raynal 1798, 7:124; Moore 2005, 29. Though most often a man of words rather than action, in this case Raynal advocated violent insurrection. Putting his own words in the mouth of a fictitious slave, Raynal dismisses the vacuous reasons used to defend the institution of slavery and suggests a mechanism for turning the tables: "Men or demons, whichever you are," Raynal's slave begins, "will you dare to justify the attempts you make against my independence, by pleading the right of the stronger[?] . . . If thou dost think thyself authorized to oppress me, because thou art stronger or more dexterous than I am, complain not if my vigorous arm shall rip up thy bosom in search of thy heart. . . . Be the victim in thy turn, and expiate the arm of having been the oppressor" (Moore 2005, 33).

52 *Journal Ecclésiastiques,* June 15, 1781.

53 Tolchard 1957, 6; Salmon 1976, 116.

54 Commager 1977, 124.

55 During his exile, Raynal traveled through England as well as Germany and Switzerland, and was permitted to return to France in 1784, on the condition that he stay out of Paris (Moore 2005, 73).

56 Raynal 1784, 5:350.

57 Ibid., 354–5.

58 Ibid., 352.

59 Ibid., 354.

60 Ibid., 357.

61 Ibid., 438.

62 See Rousseau's *Discourse on the Arts and Sciences* and *Discourse on Inequality.*

63 Raynal 1784, 5:152.

64 Ibid., 155.

65 Ibid., 179.

66 Ibid., 159–60.

67 Ibid., 159.

68 Ibid., 162.

69 Ibid., 357.

70 Raynal was influenced here by Montesquieu and Voltaire's ideal of the "Good Quaker" (Moore 2005, 7).

71 Raynal 1770, 6:376. Translated by William Tucker (Louisville, KY).

72 Ibid.

73 Raynal as cited in Boorstin 1948, 101.

74 Raynal 1784, 6:112.

75 This is true despite the fact that he and Thomas Paine got into an argument regarding the exact justification for America's breaking away from Great Britain. See *Philosophical and Political History* and Paine's "A Letter to the Abbé Raynal."

76 Indeed Raynal spends page after page listing, and then debunking, British arguments for crushing the American insurrection. (Raynal 1784, 6:167).

77 Letter from Jefferson to Robert Walsh, December 4, 1818. Carmichael, who was present

at the dinner with Raynal, goes further, noting, "In fact there was not one American present who could not have tost [*sic*] out of the windows any one or perhaps two of the rest of the company"; letter from Carmichael to Jefferson, October 15, 1787 (Library of Congress Jefferson Collection).

78 Raynal 1784, 6:125.
79 Ibid.,6:245.
80 Ibid., 6:249.

Notes to Chapter 4

1 For more on the general role of the Founding Fathers in the Enlightenment, see: Gay 1966–1969; Commager 1977.

2 Jefferson began writing what evolved into *Notes on the State of Virginia* in response to a series of queries posed by the Marquis de Barbé-Marbois. The queries were first given to James Madison's uncle, Joseph Jones, who was then in Congress. Jones appears to have passed along these queries to Jefferson. Marquis de Barbé-Marbois, Queries on Virginia (in the hand of Joseph Jones), October 1780, (Library of Congress Jefferson collection).

3 While I am focusing on the most well known voices of the American Revolution era, others too wrote in response to claims of American degeneracy. These include Benjamin Rush, Nicolas Collins, and William Barton (Chinard 1947).

4 Adams 1787.

5 Abigail Adams to Mrs. Shaw, November 21, 1786, in Adams 1840, 358–59.

6 James Madison to Thomas Jefferson, June 19, 1786; Jefferson replied on December 16, 1786. After apologizing for the delay in responding, Jefferson notes, "I thank you for your communications in Natural History" (Library of Congress Jefferson collection).

7 Madison uses Buffon's terms the "Belette & Roselet or Hermine."

8 James Madison to Thomas Jefferson, December 4, 1786 (Library of Congress Jefferson collection).

9 Porter 1986.

10 Franklin 1755. Also see Franklin's essay, *On Immigration*.

11 Thomas Jefferson to Marquis de Chastellux, June 7, 1785 (Library of Congress Jefferson collection).

12 Franklin and Buffon's science intersected in one other way, which had nothing to do with degeneracy. In 1750, Franklin wrote a series of letters to his British friend, Peter Collinson, in which he outlined an experiment to determine whether lightning was electrical in nature. These letters were communicated to the Royal Society, widely disseminated, and then, in 1752, published in French by Buffon and two of his colleagues. King Louis XV proceeded to order Buffon and his team to perform the experiment that Franklin had outlined, which they did on May 10, 1752 (Watson 1751; Isaacson 2003).

13 Chernow 2004.

14 The reference was to the French version, *Recherches philosophiques sur les Américains*.

15 Cohen 1995, 61.

16 Indeed, a cottage industry in studying "Jefferson as a man of science" has emerged in the field of American history (Boorstin 1948; Martin 1952; Greene 1984; Bedini 1990; Bedini 2002).

17 Jefferson to Pierre-Samuel Dupont, March 2, 1809 (Library of Congress Jefferson collection).

18 Ibid.

19 Ibid.

20 Bedini 1990.

21 Bedini 2002, 10.

22 Bedini 1990, 16.

23 Thomas Jefferson, July 27, 1821, Autobiography draft fragment, January 6 through July 27 (Library of Congress Jefferson collection).

24 Bedini 1990, 28.

25 Men like Francis Fauquier and George Wythe.

26 Jefferson to John Hollins, February 19, 1809 (Library of Congress Jefferson collection).

27 Thomas Jefferson to Thomas Mann Randolph, May 1, 1791; as in Martin 1952, 4.

28 Dupree 1957.

29 "The Sage of Monticello," *Niles Weekly Registry* 11, January 4, 1817. Also see Hatch 1998.

30 Thomas Jefferson to Richard Rush, April 26, 1824 (Library of Congress Jefferson collection).

31 Bedini 2002, 16.

32 Thomas Jefferson to Martha Jefferson Randolph, December 23, 1790 (Library of Congress Jefferson collection).

33 Martin 1952.

34 Jefferson's instructions to Meriwether Lewis, dated June 20, 1803; presented to Lewis June 30, 1803.

35 Jefferson [1787] 1999, 71n.

36 For information on European and South American responses to Buffon, Raynal, and de Pauw see Gerbi 1973, chaps. 4–7.

37 Jefferson [1787] 1999, 48.

38 Ibid., 59.

39 Thomas Jefferson to Marquis de Chastellux, June 7, 1785 (Library of Congress Jefferson collection).

40 Ibid.

41 Jefferson [1787] 1999, 48.

42 Shapin 1994.

43 Jefferson [1787] 1999, 68.

44 Ivaska-Robbins 2007.

45 Chinard 1947, 32–35; Echeverría 1957, 9.

46 Jefferson [1787] 1999, 56.

47 Ibid., 56.

48 Ibid., 56.

49 Ibid., 62.

50 Ibid., 69.

51 Thomas Jefferson to Marquis de Chastellux, June 7, 1785 (Library of Congress Jefferson collection).

52 Benjamin Franklin to Lorenzo Manini, November 19, 1784; as in Gerbi 1973, 240.

53 Benjamin Vaughn to Jefferson, January 26, 1787 (Library of Congress Jefferson collection).

54 Malone 1951, xxv–xxvi.

55 Chinard 1944, 152.

56 Jefferson to James Madison, January 30, 1787 (Library of Congress Jefferson collection).

57 Malone 1951, 21.

58 Jefferson to James Monroe, February 6, 1785 (Library of Congress Jefferson collection).

59 Boehm and Schwartz 1957.

60 Chinard 1947, 28.

61 Hoebel 1960; Smitten 1985.

62 Robertson 1777. Quote is from Robertson 1841, 128.

63 Robertson 1841, 128.

64 Ferguson 1768; Miller 1955. British author Sir Horace Walpole agreed, writing a friend that Buffon's theory led him to believe "genius does not seem to make great shoots there [America]. Horace Walpole to Sir Horace Mann, May 13, 1780 (Cunningham 1880, 7:365).

65 *Pennsylvania Packet*, published as *The Pennsylvania Packet or the General Advertiser,* January
11, 1780; *Massachusetts Spy*, published as *Massachusetts Spy: Or, Worcester Gazette,* January
1, 1784; *Continental Journal,* January 22, 1784.

66 Thomas Jefferson to Chastellux, June 7, 1785 (Library of Congress Jefferson collection).

67 Roger 2005.

68 Thomas Jefferson to Joseph Willard, March 24, 1789 (Library of Congress Jefferson collection).

Notes to Chapter 5

1 Jefferson to Major General William Phillips, June 25, 1779 (Library of Congress Jefferson collection).

2 On June 2, 1781, British lieutenant colonel Banastre Tarleton and his troops came within minutes of capturing Jefferson at Monticello.

3 Malone 1948, 50.

4 John Sullivan to Meshech Weare, December 27, 1780.

5 See footnotes to "Marbois' Queries Concerning Virginia" in Boyd 1950–1984, 4:167. Also John Sullivan to the Marquis de Barbé-Marbois, December 10, 1780, within which Sullivan provides "such answers to Your Queries as my time & materials would permit" (Library of Congress Jefferson collection).

6 Precisely when Jones received the queries is not known. The list was eventually expanded to twenty-three queries.

7 In a November 30, 1780, letter to Charles-François D'Anmours, Jefferson writes that he was "busily employed by Monsr. Marbois without his knowing it" (Library of Congress Jefferson collection).

8 "Marbois' Queries Concerning Virginia" in Boyd 1950–1984, 4:166–67.

9 Thomas Jefferson, July 27, 1821, autobiography draft fragment, January 6 through July 27 (Library of Congress Jefferson collection).

10 Jefferson to John Adams, May 17, 1818 (Library of Congress Jefferson collection).

11 Jefferson to Marbois, March 4, 1781 (Library of Congress Jefferson collection).

12 Jefferson to Marbois, December 20, 1781 (Library of Congress Jefferson collection).

13 There apparently was some mix-up in getting the material sent to Marbois, but he had it in hand by April 1782. See Marbois to Jefferson, January 29, 1782; Jacquelin Ambler to Jefferson, March 16, 1782; Jefferson to Marbois, March 24, 1782 (Library of Congress Jefferson collection).

14 Jefferson answered all of Marbois' questions, but reordered and restructured some of them. Although Jefferson used "productions mineral, vegetable and animal" in the table of contents, at the start of the actual chapter, he reverted back to something closer to Marbois's original language—"A Notice of the mines and other subterranaeous riches, its trees, plants, fruits & c."

15 Others who may have seen Jefferson's reply to Marbois were Thomas Walker, G. K. van Hogendorp, and Thomas Hutchings (Library of Congress Jefferson collection).

16 Thomas Jefferson, July 27, 1821, autobiography draft fragment (Library of Congress Jefferson collection).

17 Ibid.

18 Jefferson to Charles Thompson, May 21, 1784 (Library of Congress Jefferson collection).

19 Jefferson to James Madison, May 25, 1784 (Library of Congress Jefferson collection).

20 Lerch 1943.

21 Jefferson to James Madison, May 11, 1785 (Library of Congress Jefferson collection).

22 Ibid.

23 Jefferson to Chastellux, June 7, 1785 (Library of Congress Jefferson collection).

24 Thomas Jefferson, July 27, 1821, autobiography draft fragment (Library of Congress Jefferson collection). Medlin argues that Jefferson wanted a literal translation, while Morellet felt his job involved much more than rote translation (Medlin 1978).

25 Peden 1954, xix, n. 25.

26 Ibid., xxiii.

27 Ibid., xx.

28 For example, parts of *Notes* were reprinted in the *New Jersey Journal,* March 11, 1787; the *Massachusetts Gazette,* May 8, 1787; the *Massachusetts Centinel,* May 8, 1787; the *Pennsylvania Packet,* June 13, 1787; the *State Gazette of South Carolina,* July 29, 1793; the *Maryland Herald,* September 4, 1800.

29 Chinard 1944, 118; Peden 1954, xxv.

30 Jefferson to Alexander Donald, September 17, 1787 (Library of Congress Jefferson collection).

31 Jefferson to John W. Campbell, September 3, 1809 (Library of Congress Jefferson collection).

32 Jefferson to John Melish, December 10, 1814 (Library of Congress Jefferson collection).

33 Jefferson [1787] 1999, 5.

34 Ibid., 26.

35 Ibid., 32.

36 Ibid., 35.

37 Porter 1986.

38 Jefferson [1787] 1999, 43.

39 Semonin 2000.

40 Jefferson [1787] 1999, 48.

41 Ibid., 48.

42 Jefferson to Chastellux, June 7, 1785 1(Library of Congress Jefferson collection).

43 Jefferson [1787] 1999, 49.

44 Porter 1986, 18.

45 Jefferson [1787] 1999, 50.

46 Ibid.

47 Ibid., 57.

48 Ibid., 59.

49 Ibid.

50 Ibid.

51 Ibid.

52 Ibid., 61.

53 See also Query XI, "A Description of the Indians Established in That State" in Jefferson [1787] 1999. Much has been written about Jefferson's views on Indians. For more see Wallace 2001; Sheehan 1973. As we saw in the last chapter, Jefferson claimed to know little about South American Indians, and so he felt obliged not to use them as part of his response to Buffon. What Jefferson had read about South American Indians, and what he took to be accounts similar to those used in Buffon's *Natural History,* he dismissed as fairy tales akin to Aesop's fables.

54 Jefferson [1787] 1999, 63.

55 Ibid., 63–64. Even when Jefferson granted the existence of Indian traits that Buffon might have claimed were degenerate, he attributed the cause to other factors. For example, Jefferson wrote that "they raise fewer children than we do," but instead of conceding that this represented a symptom of degeneracy, à la Buffon, he argued that this was simply due to the fact that Indians relied on a hunter-gatherer lifestyle, and hence were less well fed than whites, who farmed the land to increase their food supplies. The difference between Indian and white fecundity was "to be found, not in a difference of nature, but of circumstance," Jefferson wrote in Query VI; Indian women "married to

white traders, who feed them and their children plentifully and regularly, who exempt them from excessive drudgery, who keep them stationary and unexposed to accident, produce and raise as many children as the white women," (65).

56 Ibid., 66.

57 Ibid., 67.

58 Ibid.

59 Ibid.

60 Luther Martin came to Cresap's defense in a letter published in the *Federal Gazette & Baltimore Daily Advertiser,* July 22, 1797.

61 Jefferson [1787] 1999, 67.

62 The Mingos were joined by the members of the Shawnee and Delaware tribes.

63 Jefferson [1787] 1999, 67–68. In appendix 3 of *Notes,* Jefferson provided more documentation on the history behind the story of Logan.

64 Though he thought that the Indians fared well enough in comparison to the Europeans of his day, Jefferson felt that a fairer comparison would be with whites who had not yet been introduced to some form of the written word.

65 Jefferson [1787] 1999, 68.

66 Ibid., 68.

67 Ibid., 68–69.

68 Holland 2001.

69 "Blacks" may not be the politically correct term today, but it was the way Jefferson referred to African Americans.

70 Jefferson does, however, mention albino Negroes as an "anomaly of nature." Jefferson [1787] 1999, 77.

71 Malone 1951, 101.

72 Jefferson does mention blacks in the context of natural history at the end of Query XIV.

73 Jefferson [1787] 1999, 70.

74 Jefferson to Chastellux, June 7, 1785 (Library of Congress Jefferson collection).

75 Jefferson to James Madison, May 11, 1785 (Library of Congress Jefferson collection).

76 Peden 1954, xxiv.

77 Jefferson to Samuel Brown, March 25, 1798 (Library of Congress Jefferson collection).

78 Jefferson [1787] 1999, 144–45.

79 Ibid., 145.

80 Ibid.

81 Ibid., 146.

82 Ibid.

83 Ibid.

84 Ibid., 147.

85 Jefferson initially separated differences in intelligence from differences in matters of decency: "Whether further observation will or will not verify the conjecture, that nature has been less bountiful to them in the endowments of the head," Jefferson wrote, "I believe that in those of the heart she will be found to have done them justice." Ibid., 149.

86 Ibid., 150–51.

87 Ibid., Query XIV.

88 Ibid., 69.

89 Ibid.

90 Ibid.

91 Ibid.

92 Ibid., 70. As we learned in chapter 3, in later editions Raynal subtly shifted from applying his degeneracy theory to all of America—North and South—by omitting direct

discussion of North America, and focusing on South America. Jefferson owned a 1780 edition of Raynal, and so knew of Raynal's change of heart on degeneracy and Creoles: ibid., 308, n. 112. Jefferson noted that "in a later edition of the Abbé Raynal's work, he has withdrawn his censure from that part of the new world inhabited by the Federo-Americans; but has left it still on the other parts. North America has always been more accessible to strangers than South. If he was mistaken then as to the former, he may be so as to the latter. The glimmerings which reach us from South America enable us only to see that its inhabitants are held under the accumulated pressure of slavery, superstition, and ignorance. Whenever they shall be able to rise under this weight, and to shew themselves to the rest of the world, they will probably shew they are like the rest of the world." Ibid., 71.

Notes to Chapter 6

1 Buffon 1749–1804, 9:173–88; Smellie 1781, 5:177.
2 Buffon 1749–1804, 9:220–21; Smellie 1781, 5:202–3.
3 *New York Journal*, February 12, 1767; *Boston Gazette*, July 8, 1765; *Charleston Morning Post*, February 16, 1786.
4 The ship set sail on July 5, 1784 (Malone 1948).
5 Webster 1824.
6 Jefferson to G. K. van Hogendorp, October 13, 1785 (Library of Congress Jefferson collection).
7 Count Buffon to Thomas Jefferson, December 31, 1785 (Library of Congress Jefferson collection).
8 Webster 1824, 376–77.
9 Ibid.
10 Webster notes that Jefferson was eighty-one or eighty-two when he told this story, which puts the date of the story around 1824. Jefferson also makes reference to the panther skin in a December 23, 1786, letter to Francis Hopkinson: "M. Buffon . . . did not know our panther. I gave him the stuffed skin of one I had bought in Philadelphia and it presents him a new species, which will appear in his next volume."
11 The story of the mammoth is told in Semonin 2000; Cohen 2002; Thomson 2008a.
12 Semonin 2000, 15; Weld 1848, 421.
13 Diary entry of Taylor, June 1706, as copied in 1760 by his grandson, Ezra Stiles (as in Semonin 2000).
14 Stanford 1959, 59.
15 *Boston News-Letter*, July 30, 1705.
16 Sloan 1727–1728a and b.
17 "Mammoth" first came into English usage in the 1690s (Semonin 2000, 68).
18 In 1767, the queen's physician, William Hunter, called the remains of one salt-lick specimen the "American incognitum" (Hunter 1769; Semonin 2000, 137).
19 The fossils were sent to Franklin by an Indian trader named George Croghan (Semonin 2000, 109).
20 A state park is now located at this spot. An advertisement notes, "The fossilized remains of these prehistoric animals were discovered in 1739 and displayed extensively at museums throughout the world. Notable Americans such as Thomas Jefferson and Benjamin Franklin personally examined the fossils, many of which are on display today at Big Bone Lick Museum."
21 Weld 1848, 42; Semonin 2000, 5.
22 Buffon 1749–1804, 9:126–27; Smellie 1781, 5:149.
23 Buffon 1749–1804, 11:86; Barr 1792, 7:318–19.
24 Simpson 1942.

25 The tooth was given to him by a Major Arthur Campbell (Arthur Campbell to Thomas Jefferson, November 7, 1782; Library of Congress Jefferson collection), and the thigh bone was mentioned by Ezra Stiles (diary, June 8, 1784, in Dexter 1901, 3:125).

26 Thomas Jefferson to George Rogers Clark, November 26, 1782 (Library of Congress Jefferson collection).

27 In his October 1785 letter to Hogendorp, Jefferson went out of his way to note that even after Buffon received his copy of *Notes*, he "only heard his [Buffon's] sentiments on one particular of it, that of the identity of the Mammoth & Elephant. As to this he retains his opinion that they are the same."

28 Jefferson [1787] 1999, 44.

29 Ibid., 43–44.

30 Ibid., 44.

31 Ibid., 45.

32 Ibid., 46.

33 Ibid., 47–48.

34 Ibid., 43.

35 Ibid., 55.

36 Ibid., 55–56.

37 Whitty 1912; Henline 1947; Edward 1974; Wilson 2002; Grigg 2004; Thomson 2008b.

38 Jefferson [1787] 1999, 55.

39 From the journal of Ezra Stiles, March 6–24, 1788 (in Dexter 1901, 3:3, 12).

40 Jefferson used the terms "moose" and "elk" synonymously. Technically, we now recognize the moose (*Alces alces*) of the United States as different from the elk (*Cervus elaphus*) of the United States. The European elk is also called the red deer, and to complicate matters, Europeans often refer to the American moose as an elk.

41 Thomas Jefferson to George Rogers Clark, November 26, 1782 (Library of Congress Jefferson collection).

42 These questions came from replies that John Sullivan received from John McDuffee. It is likely that Jefferson had delivered these questions directly to Sullivan, and so we are forced to use the Sullivan-McDuffee version of the questions (Boyd 1950–1984, 7:24).

43 Thomas Jefferson to John Sullivan, March 12, 1784. Attachment to letter (Library of Congress Jefferson collection).

44 This survey was actually sent by Jefferson to John Sullivan, and from Sullivan to Whipple.

45 From Boyd's commentary on Jefferson's letter to Sullivan, March 14, 1784 (Boyd 1950–1984, 7:24).

46 He would later serve as New Hampshire's governor ("president").

47 At times referred to as "Jonathon Door." John Sullivan to Thomas Jefferson, March 12, 1784 (Library of Congress Jefferson collection).

48 Dore's replies were conveyed to Isaac Hasey, who in turn sent them to Sullivan. John Sullivan to Thomas Jefferson, June 22, 1784 (Library of Congress Jefferson collection).

49 Sullivan may have been making reference to these items in a later letter to Jefferson: "The articles I procured for your excellencey are yet by me as I found that you accepted an appointment and sailed for Europe and I expected that it might be as well to let them rest unless you expected them sent by Mr. House to you in France which I can as easily do from this port as to send them from Philadelphia." John Sullivan to Thomas Jefferson, March 4, 1786 (Library of Congress Jefferson collection). It is also possible that Sullivan may have been referring to these antlers in a letter to Jefferson in which he wrote, "Perhaps you may think it is strange that I have not forwarded the articles I promised." It is unclear, from the context of the letter, what these articles were, but they may have been the antlers. John Sullivan to Thomas Jefferson, January 26, 1787 (Library of Congress Jefferson collection).

50 John Sullivan to Thomas Jefferson, June 22, 1784 (Library of Congress Jefferson collection).

51 William Whipple to Thomas Jefferson, March 15, 1784 (Library of Congress Jefferson collection).

52 Thomas Jefferson to James Madison, May 25, 1784; George Washington to Thomas Jefferson, June 2, 1784; Thomas Jefferson to John Adams, June 19, 1784; Thomas Jefferson to Benjamin Franklin, June 19, 1784 (Library of Congress Jefferson collection).

53 Thomas Jefferson to Archibald Cary, January 7, 1786 (Library of Congress Jefferson collection).

54 Thomas Jefferson to Francis Hopkinson, December 23, 1786 (Library of Congress Jefferson collection).

55 The comparisons can get a bit confusing here. In addition to believing that Buffon confused the moose and the reindeer, Jefferson also noted that Peter Kalm called the black moose a *renne*, and that Buffon's colleague Daubenton confused the *renne* with a red deer.

56 Webster 1824, 376.

57 Ibid., 376–77.

58 Thomas Jefferson to John Sullivan, January 7, 1786 (Library of Congress Jefferson collection).

59 Thomas Jefferson to William Whipple, January 7, 1786 (Library of Congress Jefferson collection).

60 Thomas Jefferson to Archibald Stuart, January 25, 1786 (Library of Congress Jefferson collection).

61 John Sullivan to Thomas Jefferson, March 4, 1786. Jefferson briefly mentions the elk in a letter: Thomas Jefferson to Francis Hopkinson, December 23, 1786 (Library of Congress Jefferson collection).

62 John Sullivan to Thomas Jefferson, January 26, 1787 (Library of Congress Jefferson collection).

63 John Sullivan to Thomas Jefferson, April 16, 1787 (Library of Congress Jefferson collection).

64 The moose is also described in a letter from Jefferson to C. Lacépède, February 24, 1803 (Malone 1951, 100).

65 Thomas Jefferson to William Stephens Smith, September 28, 1787 (Library of Congress Jefferson collection).

66 John Sullivan to Thomas Jefferson, April 16, 1787 (Library of Congress Jefferson collection).

67 Ibid.

68 Eventually the plan to borrow the money from his brother fell through because of currency exchange problems. Sullivan obtained the money from John Adams's secretary, William Stephens Smith.

69 The cost soon inflated slightly to forty-six and a half pounds; John Sullivan to Thomas Jefferson, April 27, 1787 (Library of Congress Jefferson collection).

70 John Sullivan to Thomas Jefferson, April 26, 1787 "with account of expenses for obtaining moose skeleton" (Library of Congress Jefferson collection).

71 John Sullivan to Thomas Jefferson, April 16, 1787; John Sullivan to Thomas Jefferson, April 26, 1787; Samuel Pierce to John Sullivan, April 12, 1787 (all in Library of Congress Jefferson collection). The change in plans regarding Pierce is recorded in Boyd's footnote to the April 16 letter (Boyd 1950–1984).

72 John Sullivan to Thomas Jefferson, May 9, 1787; John Sullivan to Thomas Jefferson, May 29, 1787 (Library of Congress Jefferson collection).

73 John Sullivan to Thomas Jefferson, May 9, 1787, was received by Jefferson on September 1, 1787; John Sullivan to Thomas Jefferson, May 29, 1787, was received by Jefferson

on September 26, 1787 (Library of Congress Jefferson collection; some of these dates are also recorded in the footnotes in Boyd 1950–1984).

74 Thomas Jefferson to John Sullivan August 15, 1787 (Library of Congress Jefferson collection).

75 Thomas Jefferson to William Stephens Smith, August 15, 1787 (Library of Congress Jefferson collection).

76 Thomas Jefferson to Daubenton, October 1, 1787 (Library of Congress Jefferson collection). The next day Daubenton replied to say that Jefferson's materials had arrived safely.

77 Thomas Jefferson to Count Buffon, October 1, 1787 (Library of Congress Jefferson collection).

78 Thomas Jefferson to John Sullivan October 5, 1787 (Library of Congress Jefferson collection).

79 Webster 1824, 377. Also see Jefferson to William Short, December 15, 1824, for a similar rendition (Library of Congress Jefferson collection).

80 Parton wrote that Buffon was said to have exclaimed, "I should have consulted you, Monsieur, before I published my book on natural history, and then I should have been sure of my facts" (Parton 1883).

Notes to Chapter 7

1 Semonin 2000, 294–95.

2 Martin 1952, 260.

3 And a founding member of the Tammany Society. He was also one of the founders of the American School of Geology.

4 *Biographical Dictionary of the United States Congress* (http://bioguide.congress.gov).

5 E.g., Thomas Jefferson to Marquis de Chastellux, June 7, 1785 (Library of Congress Jefferson collection).

6 Gerbi 1973, 81.

7 Pernety 1769.

8 De Pauw 1806, 34–35.

9 De Pauw 1768a, 2:5.

10 Prichard 1902; Duvernay-Bolens 1995.

11 Prichard 1902, 6–7.

12 Byron and Clerke 1767.

13 *Connecticut Gazette*, published as the *New-London Gazette*, September 1766, 3; *American Magazine; or, General Repository*, April 1769, 106.

14 De Pauw 1768a.

15 De Pauw 1806, 91.

16 Ibid., 48.

17 In 1771, Pernety responded to de Pauw's reply with yet another treatise on the flaws of the degeneracy hypothesis, entitled *Examen des Recherches philosophiques sur l'Amérique et les Américains et de la défense de cet ouvrage.*

18 Pernety was not the only one to respond directly to de Pauw. According to Church, "Major attacks were made by at least six individuals, representing three nationalities, and covering a period of forty years. These six persons were the Frenchmen Dom Pernety, Z. de Bonneville, Delisle de Sales, and Drouin de Bercy; the Italian Count Carli; and the Mexican historian Clavigero " (Church 1936, 194).

19 De Pauw 1806, viii.

20 Ibid., 48.

21 Ibid., 69.

22 Ibid., 68.

23 Ibid., 234–35.

24 Voltaire, letter of December 21, 1775 (as cited in Gerbi 1973, 153, n. 298). The Franklin-Voltaire meeting is described in Bancroft 1866, 9:499.

25 Kant, *Reflexionen zur Anthropologie*, as in Gerbi 1973, 329.

26 Kant, *On the Different Races of Man (Von des vershiedenen Racen der Menschen)*(1775), as cited in Gerbi 1973, 330.

27 Kant, *On the Use of Teleological Principles in Philosophy*, 1788, as cited in Gerbi 1973, 331.

28 Moore 1943.

29 Goldsmith 1824, 2:338.

30 Moore 1943.

31 Hegel, *Enzyklopadie der philosophischen Wissenschaften* (Encyclopedia of the Philosophical Sciences) 24, as cited in Gerbi 1973, 422.

32 Hegel, *Philosophie der Geschichte*, 1:189–91, as cited in Gerbi 1973, 426.

33 Hegel, *Philosophie der Geschichte*, 1:182, as cited in Gerbi 1973, 428.

34 Hegel, *Enzyklopadie*, 30, as cited in Gerbi 1973, 430.

35 John Keats to George and Georgina Keats, April 28, 1819 (Forman 1895); Thomas Jefferson to Chastellux, June 7, 1785 (Library of Congress Jefferson collection).

36 Briggs 1944, 187.

37 Selincourt 1905, 253. Briggs calls it "the second ode to Fannie Brawne" (Briggs 1944). It is also sometimes referred to as "What Can I do to Drive Away."

38 Stanzas 30, 31, 34, 35, 38, 39, 42, 43.

39 Briggs 1944.

40 Robertson 1777.

41 Smith 1820; Spiller 1929.

42 Lamarck even acted as a guardian and chaperone for Buffon's son at one point (Roger 1997).

43 Lamarck 1809.

44 Darwin 1845; December 23, 1833; and January 9, 1834.

45 Jefferson to Thomas Appleton, July 18, 1816 (Library of Congress Jefferson collection).

46 Marraro 1935.

47 Mazzei [1788] 1976.

48 Mazzei [1845] 1980, 294.

49 Mazzei [1845] 1980, 294.

50 Mazzei [1788] 1976, 205.

51 Ibid., 243.

52 Ibid.

53 Jefferson to Hogendorp, August 25, 1786 (Library of Congress Jefferson collection).

54 Jefferson to Hogendorp, April 4, 1787 (Library of Congress Jefferson collection).

55 Mazzei [1788] 1976, 243.

56 Byron in "Ode to Napoleon."

57 Moore 1830, 433. Byron, however, took an anti-American turn on the eve of his death (Gerbi 1973, 347, n. 84).

58 Moore 1830, 433.

59 Shelley, in *Revolt of Islam* (1817) (Brooke 1904).

60 Humboldt 1849, 119.

61 Ibid.

62 Humboldt to his brother, October 17, 1800, as cited in Gerbi 1973, 408.

63 Humboldt [1811] 1966, 3:48. In his *American Ornithology* (1808), America's first ornithologist, Alexander Wilson, agreed, and while describing the American redstart bird (*Muscicapa ruticilla*), he chided Buffon's "mistaken opinions and groundless prejudice."

64 Humboldt [1811] 1966, 1:lxxxi.

65 Ibid., 29.

66 Ibid., 2:385.

67 Humboldt 1814; Gerbi 1973, 415; Church 1936, 195.

68 Humboldt to Jefferson, May 24, 1804 (Library of Congress Jefferson collection). Also de Terra 1959.

69 Semonin 2000; Schwartz 2001.

70 Humboldt to Jefferson, June 12, 1809, and Humboldt to Jefferson, December 20, 1811 (Library of Congress Jefferson collection).

Notes to Chapter 8

1 Miller 1955.

2 His son, Samuel, would go on to develop "Morse code" (Moss 1955; Moore 1939).

3 This began as a series of lectures for young girls at a New Haven school where Morse was teaching; his friends convinced him to develop these lectures into a small book (Phillips 1983, 18–19).

4 Carpenter 1963; Phillips 1983, 34.

5 Morse 1790.

6 Ibid., 7–8.

7 Ibid., 9. Here Morse relies on comments from Abbé Clavigero.

8 Ibid., 11.

9 Ibid., 14. Other historical tracts of the period also dealt with degeneracy, but in less direct ways.

10 Cooper 1845, 50.

11 Samuel L. Mitchill, July 4, 1826 (Martin 1952, 260).

12 Thomas Grimke's Fourth of July oration of 1809, delivered in St. Paul's Church, Charleston, SC (Huddleston 1966, 316.)

13 Clinton 1815.

14 Also referred to as "the Connecticut wits."

15 Joel Barlow to Thomas Jefferson, June 15, 1787 (Library of Congress Jefferson collection).

16 Originally called "Vision of Columbus."

17 Humphreys 1804.

18 Although technically not part of the songbird family, many eagles use calls and songs as part of their mating ritual.

19 Irving 1849.

20 *The Sketch-Book of Geoffrey Crayon* was originally published as a series of essays between 1818 and 1820. Quotes from Irving 1846, 2–3.

21 Ibid., 2.

22 Ibid.

23 Ibid.

24 Pagden 1982.

25 Irving 1846, 257.

26 Ibid., 248. Indeed, it was the white settlers on the frontier, the "miserable hordes which infest," as described in *The Sketch-Book,* that were "composed of degenerate beings.

27 Ibid., 251.

28 Ibid., 248.

29 Ibid., 250.

30 Ibid., 255.

31 Ibid., 257–58.

32 Many of these essays can be found collected in Thoreau 1913.

33 Thoreau 1913, 154.

34 Ibid., 163.

35 Ibid., 156.

36 Ibid., 169–70.

37 Ibid., 173.

38 Ibid..
39 Ibid., 174.
40 Ibid., 182.
41 Ibid., 174.
42 Ibid., 175.
43 Ibid., 174–75.
44 Thoreau 1906, journal entry for January 23, 1858.
45 Ibid., April 16, 1841.
46 Exactly what Thoreau planned to do with all this data is a matter of some contention.
47 "Civilization," in Emerson 1880, 2:26. The passage begins as follows: "Climate has much to do with this melioration. Civilization is the result of highly complex organization. In the snake, all the organs are sheathed; no hands, no feet, no fins, no wings. In bird and beast, the organs are released, and begin to play. In man, they are all unbound, and full of joyful action. With this unswaddling he receives the absolute illumination we call Reason, and thereby true liberty."
48 Emerson 1880, 26.
49 Gerbi 1973.
50 "Civilization," in Emerson 1909, 10.
51 "The Poet," in Emerson 1909, 186.
52 "The American Scholar," in Emerson 1909, 21.
53 "Culture," in Emerson 1880, 2:115.
54 "The American Scholar," in Emerson 1909, 5.
55 Gordon 2004.

REFERENCE LIST

Adams, A. 1840. *Letters of Mrs. Adams*. Boston: Little Brown.

Adams, J. 1787. *A Defence of the Constitutions of Government of the United States of America*. London: C. Dilly.

Bancroft, G. 1866. *History of the United States*. Boston: Little, Brown and Company.

Barr, J. S. 1792. *Barr's Buffon*. 10 vols. London: J. S. Barr.

Barton, B. 1803. *Elements of Botany; or, Outlines of the Natural History of Vegetables*. Philadelphia.

Beall, O. T. 1961. "Cotton Mather's Early 'Curiosa Americana' and the Boston Philosophical Society of 1683." *William and Mary Quarterly* 18:360–72.

Bedini, S. 1990. *Thomas Jefferson: Statesman of Science*. New York: Macmillan.

———. 2002. *Jefferson and Science*. Virginia: Thomas Jefferson Foundation.

Beyerhaus, G. 1926. "Abbe de Pauw und Friedrich der grosse, eine Abrechnung mit Voltaire." *Historische Zeitschrift* 124:465–93.

Blanckaert, C. 1993. "Buffon and the Natural History of Man: Writing History and the Foundational Myth of Anthropology." *History of the Human Sciences* 6:13–50.

Boehm, D., and E. Schwartz. 1957. "Jefferson and the Theory of Degeneracy." *American Quarterly* 9:448–53.

Boorstin, D. 1948. *The Lost World of Thomas Jefferson*. New York: Henry Holt.

Boyd, J. P., ed. 1950–1984. *The Papers of Thomas Jefferson*. Vols. 1–20. Princeton: Princeton University Press.

Brickell, J. 1737. *The Natural History of North Carolina: With an Account of the Trade, Manners, and Customs of the Christian and Indian Inhabitants, Illustrated with Copper-Plates, Whereon Are Curiously Engraved the Map of the Country, Several Strange Beasts, Birds, Fishes, Snakes, Insects, Trees, and Plants, &c*. Dublin: James Carson.

Briggs, H. 1944. "Keats, Robertson and That Most Hated Land." *PMLA* 59:184–99.

Brooke, S., ed. 1904. *Poems of Shelley*. New York: Macmillan.

Buffon, H. N. 1860. *Correspondance inédit*.

Buffon, G.-L. 1749–1804. *Histoire Naturelle*. Paris: Imprimerie Royal, puis Plassan.

Byron, J., and C. Clerke. 1767. *Voyage Round the World, in His Majesty's Ship the Dolphin, Commanded by Commodore Byron; In Which Is Contained, a Faithful Account of the Several Places, People, Plants, Animals, &c. Seen on the Voyage*. London: Newberry.

Carpenter, C. H. 1963. *History of American Schoolbooks*. Philadelphia: University of Pennsylvania Press.

Case-Winters, A. 2000. "The Argument from Design: What Is at Stake Theologically?" *Zygon* 35, no. 1:69–81.

Chernow, R. 2004. *Alexander Hamilton*. New York: Penguin Publishers.

Chetwood, W. 1720. *The Voyages, Dangerous Adventures and Imminent Escapes of Captain Richard Falconer*. London: W. Chetwood.

Chinard, G. 1944. *Thomas Jefferson: The Apostle of Americanism*. Boston: Little Brown & Co.

———. 1947. "Eighteenth Century Theories on America as a Human Habitat." *Proceedings of the American Philosophical Society* 91:27–57.

Church, H. 1936. "Corneille De Pauw and the Controversy over His *Recherches philosophiques sur les Américains*." *PMLA* 51:178–207.

Clinton, D. 1815. *An Introductory Discourse*. New York: David Longworth.

Cohen, C. 2002. *The Fate of the Mammoth: Fossils, Myth, and History*. Translated by William Rodarmor. Chicago: University of Chicago Press.

Cohen, I. B. 1995. *Science and the Founding Fathers*. New York: W. W. Norton.

Commager, H. Steel. 1977. *Empire of Reason*. Garden City: Doubleday.

Conklin, E. 1947. "Eighteenth Century Theories on America as a Human Habitat." *Proceedings of the American Philosophical Society* 91:1–9.

Cooper, J, Fenimore. 1845. *Satanstoe*. New York: Burgess, Stringer and Company.

Cunningham, D. 1908. "Anthropology in the Eighteenth Century." *Journal of the Royal Anthropological Institute of Great Britain and Ireland* 38:10–35.

Cunningham, P., ed. 1880. *The Letters of Horace Walpole*. London: Bickers and Son.

Cuvier, G. 1800. "Notice historique sur Daubenton." *Magasin encyclopédique*: 438–69.

Danzer, G. 1974. "Has the Discovery of America Been Useful or Hurtful to Mankind? Yesterday's Questions and Today's Students." *History Teacher* 7:192–206.

Darwin, C. 1845. *The Voyage of the Beagle*. London: John Murray.

———. 1859. *On the Origin of Species*. 1st ed. London: J. Murray.

de Pauw, C. 1768a. *Recherches philosophiques sur les Américains, ou mémoires intéressants pour servir à l'histoire de l'espèce humaine; Avec une dissertation sur l'Amérique & les Américains*. Berlin.

———. 1768b. *Défense des recherches philosophiques sur les Américains*. Berlin.

———. 1806. *A General History of the Americans, or Their Customs, Manners and Colour*. Edited and translated by Daniel Webb. Rochdale: T. Wood.

de Terra, H. 1959. "Alexander Von Humboldt's Correspondence with Jefferson, Madison and Gallatin." *Proceedings of the American Philosophical Society* 103:783–806.

Dexter, F., ed. 1901. *The Literary Diary of Ezra Stiles*. New York: Scribner.

du Pratz, Antoine Simon le Page. 1758. *Histoire de la Louisiane*. Paris.

Duchet, M. 1991. "L'histoire des deux Indes: Sources et structure d'un texte polyphonique." In *Lectures de Raynal*, edited by H. J. Lusebrink and M. Tietz. Oxford: Voltaire Foundation.

Dudley, P. 1720–1721. "An Account of the Method of Making Sugar from the Juice of the Maple Tree in New England." *Philosophical Transactions of the Royal Society* 31:27–28.

Dumont, J.-F. B. 1753. *Mémoires historiques sur la Louisiane*. 2 vols. Paris.

Dupree, A. Hunter. 1957. *Science in the Federal Government*. Cambridge, MA.: The Belknap Press of Harvard University.

Duvernay-Bolens, J. 1995. *Les géants patagons: Voyage aux origines de l'homme*. Paris: Michalon.

Echeverría, D. 1957. *Mirage in the West*. Princeton: Princeton University Press.

Edward, B. C. 1974. "Jefferson, Sullivan, and the Moose." *American History Illustrated* 9:18–19.

Emerson, R. E. 1880. *The Works of Ralph Waldo Emerson*. Boston: Houghton, Osgood and Company.

———. 1909. *Essays and English Traits*. Cambridge: Harvard Classics.

Farber, P. L. 1998. "Buffon: A Life in Natural History." *Journal of the History of Biology* 31, no. 2:298–300.

———. 2000. *Finding Order in Nature: The Naturalist Tradition from Linnaeus to E. O. Wilson.* Baltimore: John Hopkins Press.

Fellows, O., and S. Milliken. 1972. *Buffon.* New York: Twayne Publishers.

Ferguson, A. 1768. *Essay on the History of Civil Society.* London: Miller and Cadell.

Ford, P. L., ed. 1984. *The Writings of Thomas Jefferson, 1781–1784.* New York: Putnam.

Forman, H. B., ed. 1895. *The Letters of John Keats.* London: Reeves and Turner.

Franklin, B. 1755. *Observations Concerning the Increase of Mankind, Peopling of Countries, &c.* Boston: S. Kneeland.

Gay, P. 1966–1969. *The Enlightenment: An Interpretation.* New York: Knopf.

Gerbi, A. 1973. *The Dispute of the New World: The History of a Polemic, 1750–1900.* Translated by J. Moyle. Pittsburgh, PA: University of Pittsburgh Press.

Gillespie, N. C. 1987. "Natural History, Natural Theology and Social Order." *Journal of the History of Biology* 20, no. 1:1–49.

Glacken, C. 1967. *Traces on a Rhodian Shore.* Berkeley: University of California Press.

Goldsmith, O. 1824. *A History of the Earth and Animated Nature.* Philadelphia: William Charlton Wright.

Gordon, J. Steele. 2004. *An Empire of Wealth: The Epic History of American Economic Power.* New York: Harper Collins.

Greene, J. 1984. *American Science in the Age of Jefferson*: Iowa State University Press.

Grigg, W. 2004. "The Moose and Thomas Jefferson." *Moose News* 8.

Grimm, F. M. 1877–1882. *Correspondance littéraire, philosophique et critique par Grimm, Diderot, Raynal, Meister.* Paris.

Hatch, P. 1998. *The Fruits and Fruit Trees of Monticello*: Charlottesville: University of Virginia Press.

Henline, R. 1947. "A Study on the *Notes on the State of Virginia* as Evidence of Jefferson's Reaction against the Theories of the French Naturalists." *Virginia Magazine of History and Biography* 55:233–46.

Hindle, B. 1956. *The Pursuit of Science in Revolutionary America, 1735–1789.* Chapel Hill: University of North Carolina Press.

Hoebel, E. 1960. "William Robertson: An 18th Century Anthologist-Historian." *American Anthropologist* 62:648–55.

Holland, C. 2001. "Notes on the State of America: Jeffersonian Democracy and the Production of a National Past." *Political Theory* 29:190–216.

Holmes, T. 1940. *Cotton Mather: A Bibliography of His Works.* Cambridge, MA: Harvard University Press.

Hornberger, T. 1935. "The Date, Source and the Significance of Cotton Mather's Interest in Science." *American Literature* 6:413–20.

Huddleston, E. L. 1966. "Topographical Poetry in the Early National Period." *American Literature* 38:303–22.

Humboldt, A. von. [1811] 1966. *Political Essay on the Kingdom of New Spain.* Translated by Mrs. Sabine. New York: AMS Press (1966).

———. 1814. *Concerning the Institutions and Monuments of the Ancient Inhabitants of America.* Translated by Mrs. Sabine. London: Longman and Hurst.

———. 1849. *Aspects of Nature in Different Lands and Different Climates.* Translated by Mrs. Sabine. Philadelphia: Lea and Blanchard.

Humphrey, J. 1898. "Manasseh Cutler." *American Naturalist* 32:75–80.

Humphreys, D. 1804. *The Miscellaneous Works of David Humphreys.* New York: T. and J. Swords.

Hunter, W. 1769. "Observations of the Bones Commonly Supposed to Be Elephant Bones Which Have Been Found near the River Ohio in America." *Philosophical Transactions of the Royal Society of London* 58:34–45.

Irving, W. 1846. *The Sketch Book of Geoffrey Crayon.* Paris: Baudry's European Library.

———. 1849. *Oliver Goldsmith: A Biography.* London: John Murray.

Isaacson, W. 2003. *Benjamin Franklin: An American Life*. New York: Simon and Schuster.

Ivaska-Robbins, P. 2007. *The Travels of Peter Kalm: Finnish-Swedish Naturalist through Colonial North America, 1748–1751*. Fleischmanns, NY: Purple Mountain Press.

Jefferson, T. [1787] 1999. *Notes on the State of Virginia*. New York: Penguin Press.

Jeske, J. 1986. "Cotton Mather: Physico-Theologian." *Journal of the History of Ideas* 47:583–94.

LaCorne, D. 2005. "Anti-Americanism and Americanophobia: A French Perspective." In *With Us or Against Us: Essays on Global Anti-Americanism*, edited by T. Judt and D. LaCorne. New York: Palgrave.

Lamarck, J.-B. 1809. *Zoological Philosophy*. Paris: Dentu.

Lerch, A. 1943. "Who Was the Printer of Jefferson's *Notes?*" In *Bookmen's Holiday*, 44–56.

Loveland, J. 2001. *Rhetoric and Natural History: Buffon in Polemical and Literary Context*. Oxford: Voltaire Foundation.

———. 2004. "George-Louis Leclerc De Buffon's *Histoire naturelle* in English, 1775–1815." *Archives of Natural History* 31:214–35.

———. 2006. "Another Daubenton, Another Histoire naturelle." *Journal of the History of Biology* 39:457–91.

Lusebrink, H. J., and A. Mussard, eds. 1994. *Avantages et désavantages de la découverte de l'Amérique*. Saint-Etienne: Publications de l'Université de Saint-Etienne.

Lusebrink, H. J., and A. Strugnell. 1995. *L'histoire des deux Indes*. Oxford: Voltaire Foundation.

Lyon, J., and P. Sloan. 1981. *From Natural History to the History of Nature: Readings from Buffon and His Critics*. Notre Dame: University of Notre Dame Press.

Malone, D. 1948. *Jefferson: The Virginian*. New York: Little Brown.

———. 1951. *Jefferson and the Rights of Man*. New York: Little Brown.

Marraro, H., ed. 1935. *Philip Mazzei, Virginia's Agent in Europe; the Story of His Mission as Related in His Own Dispatches and Other Documents*. New York: New York Public Library.

Martin, E. T. 1952. *Thomas Jefferson: Scientist*. New York: Schuman.

Masterson, J. 1946. "Traveler's Tales of Colonial Natural History." *Journal of American Folklore* 59:51–67 and 174–88.

Mather, C. 1693. *Winter Meditations*. London: B. Harris.

———. [1721] 1815. *The Christian Philosopher: A Collection of the Best Discoveries in Nature with Religious Improvement*. London: E. Matthews.

Mazzei, P. [1788] 1976. *Historical and Political Researches on the United States of America*. Translated by C. Sherman. Charlottesville: University of Virginia Press.

———. [1845] 1980. *My Life and Wanderings*. Translated by E. S. Scalia. Morristown, NJ: American Institute of Italian Studies.

Medlin, D. 1978. "Thomas Jefferson, André Morellet, and the French Version of *Notes on the State of Virginia*." *William and Mary Quarterly* 35:85–99.

Miller, R. 1955. "American Nationalism as a Theory of Nature." *William and Mary Quarterly* 12:74–95.

Moore, A. 2005. "French Observations of America: Intercultural Commentary in the Age of Revolution." PhD thesis, Catholic University of America.

Moore, J. K. 1939. *Jedidiah Morse: A Champion of New England Orthodoxy*. New York: Columbia University Press.

Moore, J. R. 1943. "Goldsmith's Degenerate Song-Birds: An Eighteenth-Century Fallacy in Ornithology." *Isis* 34:324–27.

Moore, T. 1830. *Letters and Journals of Lord Byron*. Paris: A. & W. Galignani.

Mornet, D. 1910. "Les enseignements des bibliothèques privées (1750–1780)." *Revue d'histoire littéraire de la France* 17:449–96.

Morse, J. 1790. *The History of America in Two Books*. Philadelphia: Thomas Dobson.

Moss, R. 1995. *The Life of Jedidiah Morse: A Station of Peculiar Exposure*. Knoxville: University of Tennessee Press.

Niklaus, R. 1995. "Advantages and Disadvantages of the Discovery of America, Chastellux, Raynal and the Académie-de-Lyon Competition" *Modern Language Review* 90:1009.

Pagden, A. 1982. *The Fall of Man*. London: Cambridge University Press.

Paley, W. 1802. *Natural Theology; or, Evidences of the Existence and Attributes of the Deity*. London: H. Maxwell.

Parrish, S. 2006. *American Curiosity: Cultures of Natural History in the Colonial British Atlantic World*. Chapel Hill: University of North Carolina Press.

Parton, J. 1883. *Thomas Jefferson*. Boston: Houghton Mifflin.

Peden, W. 1954. *Introduction To "Notes on the State of Virginia."* Raleigh: University of North Carolina Press.

Pernety, A. 1769. *Dissertation sur l'Amérique et les Américains contre les recherches philosophiques de M. de Pauw*. Berlin: Samuel Pitra.

Phillips, J. 1983. *Jedidiah Morse and New England Congregationalism*. New Brunswick, NJ: Rutgers University Press.

Piveteau, P. 1952. "La pensée religieuse de Buffon." *Paris Buffon Symposium*: 125–32.

Porter, C. M. 1986. *The Eagle's Nest: Natural History and American Ideas, 1812–1842*. Tuscaloosa: University of Alabama Press.

Prichard, H. H. 1902. *Through the Heart of Patagonia*. New York: D. Appleton.

Ray, J. 1717. *The Wisdom of God Manifested in the Works of the Creation, in Two Parts*. London: R. Harbin.

Raynal, G. 1770. *Histoire philosophique et politique des établissements des Européens dans les deux Indes*. 6 vols. Amsterdam.

——. 1784. *A Philosophical and Political History of the Settlements and Trade of the Europeans in the East and West Indies*. Translated by J. O. Justamond, F.R.S. 6 vols. Dublin: John Exshaw, Luke White.

——. 1798. *A Philosophical and Political History of the Settlements and Trade of the Europeans in the East and West Indies*. Translated by J. O. Justamond. 6 vols. London: Mundell & Co.

Robertson, W. 1777. *History of America*. 2 vols. Vol. 1. London: W. Strahan.

——. 1841. *History of America*. 2 vols. Vol. 1. New York: Harper and Brothers.

Roger, J. 1997. *Buffon: A Life in Natural History*. Translated by S. Bonnefoi. Ithaca: Cornell University Press.

Roger, P. 2005. *The American Enemy: The History of French Anti-Americanism*. Translated by S. Bowman. Chicago: University of Chicago Press.

Roland, M.-J. 1976. *Private Memoirs of Madame Roland*. New York: ASM Press.

Ruse, M. 2003. *Darwin and Design*. Cambridge: Harvard University Press.

Salmon, J. H. M. 1976. "Raynal, 1713–1796: An Intellectual Odyssey." *History Today* 26, no. 2:109–17.

Schwartz, I. 2001. "Alexander von Humboldt's Visit to Washington and Philadelphia, His Friendship with Jefferson and His Fascination with the United States." *Northwest Naturalist* special issue 1:43–56.

Selincourt, E. 1905. *The Poems of John Keats*. New York: Dodd, Mead and Company.

Semonin, P. 2000. *American Monster: How the Nation's First Prehistoric Creature Became a Symbol of National Identity*. New York: New York University Press.

Shapin, S. 1994. *A Social History of Truth, Civility and Science in Seventeenth-Century England*. Chicago: University of Chicago Press.

Sheehan, R. 1973. *Seeds of Extinction: Jeffersonian Philanthropy and the American Indian*. Chapel Hill: University of North Carolina Press.

Simpson, G. G. 1942. "The Beginnings of Vertebrate Paleontology in North America." *Proceedings of the American Philosophical Society* 86:130–88.

Sloan, H. 1727–1728a. "An Account of Elephants Teeth and Bones Found Underground." *Philosophical Transactions of the Royal Society of London* 35:457–71.

——. 1727–1728b. "Of Fossil Teeth and Bones of Elephants, Part the Second." *Philosophical Transactions of the Royal Society of London* 35:497–514.

Smellie, W., trans. 1781. *Buffon's Natural History*. 8 vols. London: Strahan and Cadell.

Smith, S. 1820. "Review of Statistical Annals of the United States of America." *Edinburgh Review* 33:69–80.

Smitten, J. 1985. "Impartiality in Robertson's History of America." *Eighteenth Century Studies* 19:56–77.

Sorel, A. 1897. *L'Europe et la révolution française*. Paris.

Spiller, R. E. 1929. "The Verdict of Sidney Smith." *American Literature* 1:3–13.

Stanford, D. 1959. "The Giant Bones of Claverack, New York, 1705, Described by the Colonial Poet, Reverend Edward Taylor (ca. 1642–1729) in a Manuscript Owned by Yale University." *New York History* 40:47–61.

Thomson, K. S. 2008a. "Jefferson, Buffon, and the Moose." *American Scientist* 96, no. 3.

———. 2008b. *The Legacy of the Mastodon: The Golden Age of Fossils in America*. New Haven, CT: Yale University Press.

Thoreau, H. D. 1906. *The Writings of Henry David Thoreau*. Boston: Houghton-Mifflin.

———. 1913. *Excursions*. New York: Thomas Crowell Company.

Tolchard, V. R. 1957. "Abbé Raynal: Advocate of Revolution and Reaction." MA thesis, University of California.

Topinard, P. 1883. "Buffon, anthropologiste." *Revue d'Anthropologie* 6:35–55.

Wallace, A. F. 2001. *Jefferson and the Indians: The Tragic Fate of the First Americans*. Cambridge, MA: Belknap Press.

Watson, W. 1751. "An Account of Mr. Benjamin Franklin's Treatise, Lately Published, Instituted Experiments and Observation on Electricity, Made at Philadelphia in America." *Philosophical Transactions of the Royal Society*.

Webster, D. 1824. "Notes on Mr. Jefferson's Conversation 1824, at Monticello, 1825." In *The Papers of Daniel Webster: Correspondence, 1798–1824*. Vol. 1. Edited by C. Wiltse and H. Moser (Hanover, NH: Dartmouth College/University Press of New England, 1974).

Weld, C. R. 1848. *A History of the Royal Society*. London: John W. Parker.

Whitty, J. 1912. "Thomas Jefferson's Bull Moose." *Nation* 95:211.

Wilson, G. 2002. "Jefferson, Buffon, and the Mighty American Moose." *Monticello Newsletter* 13, no. 1.

Wiltse, C., and H. Moser, eds. 1974. *The Papers of Daniel Webster: Correspondence, 1798–1824*. Vol. 1. Hanover, NH: Dartmouth College/University Press of New England.

Wolpe, H. 1957. *Raynal et sa machine de guerre*. Stanford: Stanford University Press.

Womack, W. 1970. "Eighteenth Century Themes in the *Histoire philosophique et politique des deux Indes* of Guillaume Raynal." MA thesis, University of Oklahoma.

Wood, G. S. 1982. "The Bigger the Beast the Better: Buffon's American Environmental Conclusions." *American History Illustrated* 17, no. 8:30–37.

INDEX

Page numbers in italics indicate an illustration. Page numbers with a t *indicate a table.*